Military Finances

Military Life

Military Life is a series of books for service members and their families who must deal with the significant yet often overlooked difficulties unique to life in the military. Each of the titles in the series is a comprehensive presentation of the problems that arise, solutions to these problems, and resources that are of much further help. The authors of these books—who are themselves military members and experienced writers—have personally faced these challenging situations, and understand the many complications that accompany them. This is the first stop for members of the military and their loved ones in search of information on navigating the complex world of military life.

Titles in the Series

The Wounded Warrior Handbook: A Resource Guide for Returning Veterans by Don Philpott and Janelle Hill (2008).
The Military Marriage Manual: Tactics for Successful Relationships by Janelle Hill, Cheryl Lawhorne, and Don Philpott (2010).
Combat-Related Traumatic Brain Injury and PTSD: A Resource and Recovery Guide by Cheryl Lawhorne and Don Philpott (2010).
Special Needs Families in the Military: A Resource Guide by Janelle Hill and Don Philpott (2010).
Life After the Military: A Handbook for Transitioning Veterans by Janelle Hill, Cheryl Lawhorne, and Don Philpott (2011).
Military Mental Health Care: A Guide for Service Members, Veterans, Families, and Community by Cheryl Lawhorne-Scott and Don Philpott (2013).
Military Finances: Personal Money Management for Service Members, Veterans, and Their Families by Cheryl Lawhorne-Scott and Don Philpott (2013).

Military Finances

Personal Money Management for Service Members, Veterans, and Their Families

Cheryl Lawhorne-Scott and Don Philpott

ROWMAN & LITTLEFIELD PUBLISHERS
Lanham • Boulder • New York • London

Published by Rowman & Littlefield Publishers
A wholly owned subsidiary of The Rowman & Littlefield Publishing Group, Inc.
4501 Forbes Boulevard, Suite 200, Lanham, Maryland 20706
www.rowman.com

Unit A, Whitacre Mews, 26-34 Stannary Street, London SE11 4AB

First paperback edition 2015

British Library Cataloguing in Publication Information Available

Library of Congress Cataloging-in-Publication Data

Lawhorne-Scott, Cheryl, 1968–
Military finances : personal money management for service members, veterans, and their families / Cheryl Lawhorne-Scott and Don Philpott.
pages cm — (Military life)
Includes index.
1. Soldiers—United States—Finance, Personal. 2. United States—Armed Forces—Military life—Handbooks, manuals, etc. 3. Sailors—United States—Finance, Personal. 4. Military dependents—United States—Finance, Personal. 5. Finance, Personal. I. Philpott, Don, 1946– II. Title.
UC74.L39 2013
332.0240088'35500973—dc23
2013005225

ISBN: 978-1-4422-2214-4 (cloth : alk. paper)
ISBN: 978-1-4422-5686-6 (pbk : alk. paper)
ISBN: 978-1-4422-2213-7(electronic)

∞™ The paper used in this publication meets the minimum requirements of American National Standard for Information Sciences Permanence of Paper for Printed Library Materials, ANSI/NISO Z39.48-1992.

Printed in the United States of America

Contents

Foreword

Our service members and their families are the backbone of the defense of our nation; they make real sacrifices in their lives to provide the defense that makes our nation secure. Keeping our service members and families healthy on all levels, including financial stability, is in the best interests of our nation. Financial hardships often occur in the military and can be a serious degrader to an individual and a unit's readiness. Financial difficulty affects job performance, health, marriages, friendships, and, in some cases, even job security. In my twenty-six years of service from a single enlisted man to a married officer who has counseled many Marines experiencing personal hardships, financial trouble is one of the most, if not *the* most, common hardship I have seen, and in some cases it is the reason behind other difficulties in people's lives.

Where do our service members learn about personal finances? From mentors in their unit? In boot camp? From high school or college? From parents? Maybe, maybe not at all, and if so, was it adequate? Did it properly prepare a service member or spouse for the military? Having come through the ranks starting as enlisted, earning a degree in finance, becoming an officer, getting an MBA, and then becoming a commanding officer, I will say that we inadequately prepare our service members to manage their personal finances outside of the parenting they received. Unless leaders and mentors impart proper financial guidance or the service member and/or family member seeks information or assistance, they are left to figure it out on their own. As the saying goes, an ounce of prevention is worth a pound of cure, and in many cases the request for information or help comes later than is optimal. Quite frequently, some knowledge by our service and family members and our leaders and mentors can help prevent many issues that we see.

We train our service members to fight our country's battles and win them, we train them on how to deploy, we offer assistance with deployments, and we have programs in place for support. Where is the owner's manual/guidebook for a service member? How does a leader or mentor learn to teach the service member what he or she needs to know? Who is responsible for ensuring the family members get the information they need? Leaders do their best to do right by service members and their families. In some cases someone in a unit is an expert in a given area and can help our service members and families, in some cases it is a base or station support service that we can send them to for assistance, and in some cases there are nonprofit organizations and legitimate nongovernment organizations that can help. In every case, it helps as an individual service member, as a family member, and as a leader to learn about the various resources that are available, what they can do, what is vetted as legitimate, and who is eligible to use those resources.

Regardless of whether a service member has learned something or if they haven't been taught a thing about personal finances, there are benefits and resources designed to help them, whether single or married. There are also many benefits and resources to help military spouses and children of service members. These resources are built into a service member's compensation; it is part of what we provide for them beyond just a paycheck. A resource designed to help is only good if someone knows it exists and how to access that resource. Sometimes it seems there is an overload of information out there and we hear about programs but not the ones we need. Sometimes we don't know who to ask, and sometimes those we ask don't know all that is available to them. I encourage those I lead and their families to get help, in any area they need. That help is there for a reason and that reason is to help the lives of our service members and their families.

Having a guide, an owner's manual of sorts, to navigate some of those resources and considerations with personal finances for military members would have been invaluable to me during the many years of service both as an individual and as a leader. The author, my wife, teammate, and best friend, has answered that need in this book. She has provided this resource as an easy-to-understand source of information and a guide to comprehend the resources available to service members and their families. This is a great resource for individuals, family members, and leaders.

How we take care of our service members as a nation is a reflection on the nation as a whole. It is incumbent on each of us to make sure we are getting our service members and ourselves all of the help and support that is needed.

Jeff Scott
Lieutenant Colonel, United States Marine Corps
Commanding Officer, Marine Fighter Attack Squadron 121

Acknowledgments

As with all the titles in the Military Life series, the aim is to produce a one-stop guide that, hopefully, covers all the information you need on a specific subject. We are not trying to reinvent the wheel, simply to gather information from as many available sources as possible so that you don't have to. Almost all the information in this book comes from federal and military websites and is in the public domain. These include the Department of Defense, American Forces Press Service, U.S. Army Medical Department, Department of Veterans Affairs, Department of Health and Human Services, and the websites of all branches of the U.S. military. We have tried to extract the essentials. Benefits, tax laws and regulations are constantly changing. Before making any decisions pleasure ensure you have the latest information available. Where more information might be useful, we have provided websites and a list of resources that can help you.

Introduction

Members of the military, especially those with families, often have a tough time handling their finances, both while serving and when returning to civilian life. Many serving military are young with little or no training in financial affairs, and they face frequent deployments and relocations—all of which can wreak havoc with their finances.

According to the Department of Defense, the average junior enlisted member with less than four years' experience earns just over $40,000 a year, including housing and food allowances. Service members with families earn about 25 percent a year more. Even when the serving member and spouse are both employed, it is often a struggle to find affordable housing and child care—and save for the future.

When service members leave the military either to retire or to transition into civilian employment, the situation can be even more fraught with difficulty because of the chaotic housing market and high unemployment. While ex-military can take advantage of many special benefits, it is of little comfort if there are no jobs to be had.

Under these circumstances it is difficult to sit down and try to plan for your future when you don't know how you are going to put food on your table tomorrow. Hopefully, this handbook will help provide some of the answers. No matter how much—or how little—money you have, you have to spend it wisely. You have to be financially responsible, and that means being disciplined.

You must sit down and take stock of your situation. How much do I earn? How much do I need to spend? (Remember that there is a huge difference between what you spend and what you need to spend.) What are the areas where I can cut back on spending? Until you have the answers to these questions, you cannot develop a budget of living within your means—and that budget is the road map that will guide you to where you want to be in the future.

Even if things look bleak, there are organizations that can help you. There are grants and loans for further education and assistance for all sorts of life's essentials from housing and relocation grants to help with child care, job hunting, and mobility. For our wounded warriors, there is even more help available.

Personal financial management is not difficult once you understand the basics, but it is essential if you want to plan for your future and achieve the goals you have set for yourself and your family.

Disclaimer: This book provides general guidelines on a wide range of financial issues applicable to members of the military and their families. However, it is not intended that you act solely on the basis of the information provided herein. Before making any decision affecting your finances, seek the help of a professional financial adviser, banker, or base counselor who specializes in this area.

Chapter One

Conducting Your Financial Audit

Potentially career-impacting financial issues are among the top concerns for service members and their families, according to military financial experts. At a Financial Fitness Forum in Washington, D.C., in December 2011, sponsored by the Consumer Financial Protection Bureau's Office of Servicemember Affairs, director Holly Petraeus said, "For military personnel, the consequences of a bad credit report can be devastating" and may lead to security clearance loss or, in the worst-case scenario, be a career ender.

Financial problems, she explained, are now the number one cause for security clearance loss, which may bar troops from doing their jobs. It's a roadblock, she added, that potentially could lead to separation from service.

Petraeus cited the housing market as one of the key factors causing military families financial heartache. Housing values have dropped across the nation, she noted, and some families are finding themselves stuck with a house that's worth less than what they owe on it.

Once they get orders to move, she added, they really get in a bind. They can't sell the house and pay off the mortgage due to its lessened value, and they may not be able to rent it out for enough to cover their payment. And if they're not delinquent on their home, they're unable to access various foreclosure prevention programs.

Petraeus said it's the aim of her office to offer service members and their families support as they confront these types of issues. Her office, she explained, has three primary missions: to educate and empower service members and their families to make better financial decisions, to monitor consumer complaints and subsequent responses, and to coordinate federal and state agencies' efforts to improve consumer financial protection measures.

Petraeus grew up in a military family and her husband is retired Army general and former CIA director David H. Petraeus.

Service members and their families today face the same challenges as most Americans: getting by in a tough economy, housing market struggles, and personal financial management issues.

This handbook aims to help you navigate through the complex world of finance, taxation, and investment in order to plan, hopefully, for a stress-free and financially secure future.

PLANNING FOR YOUR FUTURE

Sit down and work out a financial road map starting with where you are now and where you want to be in the future. What are your current finances like? Are you able to pay all your bills on time? Do you have credit card debt? Do you want to buy a house? Where do you want to be in ten, twenty, or thirty years? Do you want to have money set aside for the children's college fund? Do you want a retirement nest egg? When you have your answers to these questions, work out a realistic time frame in order to achieve this. How much time you have to do this will determine how much you have to save every week or month to make it happen.

First, write down all your household expenses. Tracking and managing your expenses is crucial to financial readiness, and using a budget to do so is a smart move. Learn the definitions of income and outgoings and know how much you are contributing to each category.

Your earnings from your job and interest from your savings accounts are your income. These earnings are subject to income taxes and governmental withholdings depending on the amount of income and regularity of pay (monthly, weekly, etc.) as well as your dependency status. Anything you spend your income on are outgoings, or expenses. Examples of outgoings include the following:

- *Home expenses*: Rent, mortgage, homeowners/renters insurance, property taxes, home repairs/maintenance/improvements
- *Utilities*: Electricity, gas, telephone and Internet, water and sewer
- *Food*: Groceries, eating out
- *Family obligations*: Child support, alimony, day care, babysitting
- *Health and medical*: Insurance, doctor's office/Rx copays, fitness memberships
- *Transportation*: Car payments, gasoline, auto insurance, car repair and maintenance, public transportation costs
- *Debt payments*: Credit cards, student loans, other loans
- *Entertainment and recreation*: Cable TV, videos, movies, hobbies, subscriptions and club dues, vacation expenses

- *Investments/savings*: 401(k) or IRA, stocks, bonds, mutual funds, college funds, savings, emergency fund
- *Miscellaneous*: Toiletries, clothing, household products, gifts, donations, haircuts, and others

Compare that with how much is coming in. Do you have any money left over to save? If not, where can cuts be made? What changes can you make to your lifestyle to reduce spending in order to achieve your long-term goals?

That is where budgeting comes in. Budgeting is not just managing your money but introducing ways to make it go further. Cut coupons and use them—it can add up to grocery savings of hundreds of dollars over a year. Commissaries overseas will even accept coupons up to six months after their expiration date. You can join a discount warehouse and buy in bulk. You may not want to buy forty-eight rolls of toilet paper at a time, so split your shopping with friends so that you all save.

Pay your bills early rather than waiting until the last minute and running the risk of incurring late penalties.

Another great way of budgeting is to plan your week; decide what meals you are going to eat each day, and only buy those items when you do your weekly shopping. This prevents impulse purchases and the temptation to eat out. Packed lunches are cheaper than buying lunch out. Remember, it is all the small savings that add up to the big savings.

Another important element of budgeting is to ensure that both you and your spouse have your own spending money to do with as you wish. This does not have to be a huge amount, but you should agree between you how much you each get—and then how you spend it is up to you.

Sound financial management is an ongoing process. Revisit your financial strategy every year or so. Are you still on track? Have your circumstances changed? If so, how has this impacted your budget planning, and what changes need to be made to get you back on track?

If you still have issues over finances, you should seek help. In the first instance you can speak to a financial adviser on base, your bank manager or tax preparer, or go to http://www.consumer.gov and click on "Managing Your Money," where you will find advice and publications you can download on a wide range of financial issues. If you still need help, there are many resources you can turn to, from financial and marriage counselors to nonprofit financial and credit counseling organizations.

CREATING A BUDGET

Budgeting is a method of tracking your income and expenses. By establishing a budget, you can develop a plan to effectively manage your money. This doesn't mean you will no longer be able to have fun or that you will be restricted from spending money. A budget simply allows you to take control of your financial life.

Having a budget provides knowledge of how much money you have coming in and going out and how much money is needed to maintain your lifestyle. With more knowledge, you are better able to simplify money management and be more successful in the long term.

You can create a budget on paper, online, or by using a spreadsheet. Providing as much detailed information as possible will ensure your success with budgeting. Creating a budget is very straightforward and can be accomplished by following a few easy steps:

1. *Gather financial documents* such as paychecks and other income sources, bank statements, investment accounts, recent utility bills, mortgage or rent statements, insurance, loans, and any information regarding expenses (groceries, gasoline, clothing, etc.).
2. *List your sources of income* for one month. If you are self-employed or an independent contractor, or your income varies, make a conservative estimate. Remember to include any extra money you make on tips, odd jobs, or gifts.
3. *List all of your monthly expenses* by dividing them into two categories: fixed and variable. Fixed expenses stay relatively the same each month and are required parts of your way of living. They include expenses such as your mortgage or rent, car payments, utilities, investments, loan payments, insurance (health, life, auto, disability), income taxes, college and retirement savings, and so forth. Variable expenses are likely to fluctuate each month and include items such as groceries, gasoline, travel, credit card payments, entertainment, eating out, and gifts, to name a few. This category will be important when making adjustments to your budget.
4. *Analyze the results* between your income and expenses. If your end result shows more income than expenses, you are already successful at money management. This means you can prioritize this excess to areas of your budget such as retirement savings or paying more on credit cards to eliminate that debt. If you are in a situation where the expenses exceed the income, then you will have to make changes to reduce expenses.
5. *Estimate how much money is needed for spending money*. By including this in your budget, you can avoid spending all of your extra income on random purchases. Consider saving or investing extra income to have more money to spend in the future.
6. *List goals*, then determine how much money you need to save monthly to reach these goals. Goals can be both short and long term. Setting

aside money in your budget will help you be successful in reaching these goals.

7. *Set up your budget to pay off debt.* Credit card debt, which incurs late fees and interest, is considered bad debt, so getting credit cards to a zero balance should be a priority. Loans that pay for something that retains value, such as a home or education, are considered good debt.
8. *Set aside money for the unexpected.* You should always have available funds on hand for unexpected expenses or emergencies. These funds can be built up over time but should be included in your budget. There is no specific amount you should have as a buffer, but many people try to budget an amount equal to three to six months of living expenses in case they become unemployed.
9. *Save money for retirement* to ensure you have enough money to cover expenses after you stop working. Check into employer-sponsored retirement savings plans first, as these programs may offer matching contributions from your employer. You may also consider an Individual Retirement Account (IRA) as a method of saving for your retirement income.
10. *Review your budget.* A periodic review of your budget will help you stay on track by showing where you budgeted correctly as well as revealing areas where change may be needed.

You are likely to discover that your financial life is less stressful because budgeting puts you in control of your money. You can make adjustments any time and become more aware of how you spend your money.

Budgeting and Planning

Budgeting and planning are the best keys to ensure you and your family avoid financial troubles. Very briefly, there are a few steps you can take to quickly determine your current financial situation and when you transition from the military.

1. The first step is to list your income and determine where your monetary sources come from.
2. Next, write down a list of all your expenses. Any expense which leaves your accounts should be listed. Try to remember everything from toiletries to entertainment expenses and even child care.

3. Establish a monthly budget by subtracting your total monthly expenses from your total monthly income. If you have more income than expenses, you can determine how much you want to put away for emergency spending.
4. If you find that your expenses appear to exceed your total income, you may need to rate your expenses by listing them in order of highest need, such as food, shelter, and clothing, to low needs, such as cable TV or piano lessons.
5. Once you have done this, it will be up to you to stick to your budget and work through your month-to-month expenditures. However, if you find this difficult, there are resources to assist you further. It is best to talk to financial counselors or even get free credit counseling to help you stay on top of your financial responsibilities.

Remember, it's your future. Make a commitment to yourself by mapping out your financial independence.

Money Management Tips

Live within Your Means

Although this sounds like a logical statement, living within your means can be difficult without a good financial plan and budget to get you started. Spending less than you earn is difficult if you don't look at where you're spending your money on a daily basis. Some of us don't give a second thought to going out and buying coffee in the morning. A specialty coffee drink costs around $4. Over the period of one month, if you did that three times a week, you would spend $48 for the month on coffee. In a year, that can add up to $576! A few dollars here and there can add up to a lot, so it's good to know what your expenditures are and determine if you need to cut back on your spending in some areas.

Build a Safety Net

Life happens! Will you be prepared when life brings you the unexpected? There's never going to be a perfect time to start saving. The important thing is to start, and you don't need lots of cash to do so. Having a "rainy day" or emergency fund can protect you from having to use high-cost credit when you hit a rough patch.

Most financial planners agree that having at least three to six months of living expenses in a savings account is a good idea. This may be more of an ideal situation, and not something everyone can do right away. Regardless, working toward the goal of having this money put away for emergencies, loss of a job, or a medical crisis is a good idea. If this seems like a steep goal, start small and work toward saving a little each month. This will make the goal seem more attainable. You and your family can start now by setting a savings goal and looking to reach that goal as quickly as possible.

Save for the Future

Saving for the future is different than having a safety net in case of emergencies. Sometimes people mistake one for the other, but saving for the future means working to put money away so that it's there for you when you've retired. Saving for the future is one of the harder things to do. There always seems to be some valid reason why it should be done later. Most times, it's because we're not familiar with what steps we need to take in planning for the future. Talk to a financial planner about setting up an Individual Retirement Account (IRA) or a Thrift Savings Plan (if you're a federal employee). Resources are out there to help you make the decision you need for your future. Don't delay.

Use Credit Wisely

Too many times people are taken by surprise when they find out they are not eligible for programs, employment, or credit because of what is on their credit report. It is important to request and review your credit report on a periodic basis. This will prevent you from being caught off guard and from ultimately being denied credit or a job based on a poor credit history. Your report can be reviewed to determine if you are responsible and pay your bills on time.

Everything from late utility payments to revolving credit, such as credit cards, can impact your report. For more information about your credit report and questions on building positive credit, speak to your financial planner.

GETTING AN EDUCATION

Paying for your education requires planning. The type of education you choose will greatly affect the costs. Your service in the armed forces may help reduce those costs. Eligibility and benefits vary by program. Visit each program link for more detailed information.

The Montgomery GI Bill

The MGIB Active Duty provides up to thirty-six months of educational benefits to active-duty veterans. Reservists may be eligible for the MGIB Selected Reserve.

Post-9/11 GI Bill

The Post-9/11 GI Bill provides up to thirty-six months of support for education and housing to individuals with at least ninety days of aggregate service on or after September 11, 2001, or individuals discharged with a service-oriented disability after at least thirty days.

VEAP/REAP

The Veterans Educational Assistance Program and the Reserve Educational Assistance Program are contribution-based programs to provide service members with educational funding.

Loans and Grants

There are three kinds of federal aid: grants, work-study, and student loans. Federal student aid is need-based. To find out if you are eligible, you must complete the Free Application for Federal Student Aid.

Some states offer additional educational benefits to members of the military or their family members.

There are also private loans, which are usually at a higher interest rate than the government loans. These loans do not have the same terms as federal student loans and the repayment terms may be significantly different.

Scholarships

Scholarships for military personnel and family members are available from a variety of sources, such as military professional organizations and veterans service organizations. Some are need-based; others are awarded on merit. The National Resource Directory lists a wide variety of scholarships available. The American Legion's *Need A Lift?* publication also lists a wealth of other resources to help further your education.

PROTECTING YOUR FINANCES

Service members and their families face unique financial challenges. But there are also unique resources to protect their finances and their rights.

Don't Get Ripped Off: Be Smart about Accessing Your VA Benefits

Thinking about applying for one or more of your VA benefits? Alarms should sound if you receive a call or personal visit at your door offering personalized assistance in filing for your VA benefits.

The Setup

A knock at the door or a telephone call offers military personnel direct, personalized assistance in applying for their VA benefits.

The Red Flag

The VA doesn't generally make house calls, and it doesn't participate in telemarketing.

The Scam

These people are not at your door to provide a public service or reward you for your military service. *They want your personal information and access to your financial accounts!*

The Solution

Information and access to all your VA benefits are available online through the Department of Veterans Affairs. You can also obtain your Certificate of Eligibility for your VA Home Loan Guaranty through the http://www.ebenefits.va.gov portal. All military personnel and veterans can register for access to a variety of information to assist in understanding all your benefits.

If you don't register until after you separate from the armed services, you may need to visit a VA service location to validate your credentials before registering. If you register while you are serving, you can save a trip and keep the same account for life.

The Dodd-Frank Wall Street Reform and Consumer Protection Act

This act has many provisions relevant to military members and their families.

- The law restricts prepayment penalties for mortgage loans.
- It authorizes individuals to receive their credit score free annually.
- It establishes the Consumer Financial Protection Bureau and the Office of Service Member Affairs.
- It mandates consumer financial literacy.
- It requires greater disclosure on credit card accounts and statements.
- It establishes a national consumer complaint hotline.
- It limits to $10 the maximum amount a business may set as a minimum purchase for credit card use.
- It permanently raises the FDIC limit of protection to $250,000.

- It allows for emergency homeowner loans up to $50,000 to prevent foreclosures.
- It establishes a student loan ombudsman.

PAY AND GENERAL MONEY MANAGEMENT

This section is a wide-ranging overview of areas you need to take into consideration whether you are staying in the military, transitioning into civilian life, or planning to retire. All areas are dealt with in greater detail later in this book.

Separation Pay

Separation pay is a topic you need to know about, especially if you are being involuntarily separated or released from active duty. Just knowing that separation pay may be available to you can assist in any loss of monies that may be owed to you.

Your eligibility and the amount you receive under this program are determined by the type of separation you receive. For example, if you are involuntarily or voluntarily separated under Variable Separation Incentive (VSI) or Special Separation Benefit (SSB), this could qualify you for separation pay. Unfortunately, it is not quite as simple as that. In order to establish whether you are eligible and what amount you qualify for, you will need to know your Separation Program Designator (SPD) code too. On your DD Form 214, your SPD code is found in block 26, and the narrative description for the code is located in block 28.

Primarily, there are two types of pay: full separation pay and half separation pay.

Full Separation Pay

To qualify for full pay, at a minimum, you must be involuntarily separated, have been on active duty for six years but less than twenty years, be fully qualified for retention, and the service discharge must be characterized as "Honorable." An example would be separation due to reduction in force, or separation due to exceeding the high year of tenure.

Half Separation Pay

To qualify for half pay, you must be involuntarily separated, with service characterized as "Honorable" or "General (under honorable conditions)" and the reason for discharge must be under a specified category. An example would be discharge due to failing fitness/weight standards.

Final Pay

There is also the possibility of receiving a final pay at separation that would include any earned entitlements and pay to include any accumulated leave. While this might sound great, keep in mind that the amount you receive will be reduced by taxes, any outstanding balances on advance pay you may have received, along with unearned bonus recoupment and other debts you've incurred throughout your service, such as a tuition payback.

Sometimes we forget that financial transactions may already have been in process. If so, this could potentially create a debt where there was none before. You are ultimately responsible for repayment of all debts owed to the federal government. Since the actual amount of separation pay is computed by the local finance office, specific questions you may have about SPD codes and separation pay issues should be directed to your military personnel office. For more information about separation pay, you may visit:

http://militarypay.defense.gov/benefits/Separation%20pay.html

Unemployment Compensation

As you are getting ready to separate, you may have looked into unemployment benefits. You may have thought about applying for unemployment compensation but did not know where to start. Unemployment is viewed differently by many people. Some see unemployment as a free handout from the government, while others believe that it should be a last resort. Regardless of how you feel about it, you should know what your rights are, know what you are entitled to, and find out if you are eligible.

What most service members fail to realize is that you have earned your unemployment rights. However, having earned this unemployment compensation does not necessarily make you eligible. Regardless, if you are separating from active duty and unable to find a new job, this compensation is something you need to consider.

You may be eligible if you were:

on active duty within a branch of the U.S. military
separated under honorable conditions
not making payroll deductions for unemployment insurance protection

Unemployment compensation for ex-service members is referred to as UCX. Basically an unemployment program designed to cater to the military, the program takes care of service members when they separate from active service, regardless of the state to which they return. Different states' rules and requirements for eligibility may vary slightly. You will need to file a claim in your state and from there the state workforce agency will determine your eligibility. Keep in mind that each state and its laws determine the actual amount of benefits and the number of weeks that the benefit can be paid to you.

All state laws stipulate that you must be both able and available for work. You must meet these conditions continually to receive benefits. Only minor variations exist in state laws setting forth the requirements concerning "ability to work." A few states specify that a claimant must be mentally and physically able to work.

"Available for work" means being ready, willing, and able to work. In addition to registering for work at a local employment office, most state laws require that you seek work actively or make a reasonable effort to obtain work. Generally, you may not refuse an offer of or referral to suitable work without good cause. As the length of unemployment increases, you may be required to accept a wider range of jobs.

Disqualifications

The major causes for disqualification from benefits are not being able to work or available for work, voluntary separation from work without good cause, discharge for misconduct connected with the work, refusal of suitable work without good cause, and unemployment resulting from a labor dispute. Disqualification for one of these reasons may result in a postponement of benefits for some prescribed period, a cancellation of benefit rights, or a reduction of benefits otherwise payable.

Receiving unemployment while you are looking for work will help pay bills and provide you some temporary financial support. If you are interested in learning more about unemployment compensation, go to http://www.servicelocator.org/ for more information.

Financial Management during Transition

Are you aware of the financial resources that are available during your transition? Even if you have a career lined up, chances are that you might be on a limited income or working on a new family budget. If you don't have a job lined up yet, finding one may take a little longer than you anticipate.

The Financial Services Center (FSC) offers courses and access to counselors who can assist you with some financial management planning. Whether you are on active duty, in the guard, or a reservist, the FSC has something to offer everyone in the way of financial planning.

Comparison of Military versus Civilian Pay and Benefits

In the past few years, different agencies have been tasked by the government to make a comparison between the pay and benefits of military and civilians. The National Defense Authorization Act for fiscal year 2010 required that the U.S. Government Accountability Office conduct a study comparing the pay and benefits provided to members of the armed forces with those of civilians. And although this study was to focus on how pay and benefits affected recruitment and retention in the armed forces, the findings showed some similarities in pay and benefits between the two.

Regardless of the similarities, the reality is that the type of job or career you have will greatly impact your salary and benefits. Remember, it's not just how much you make that will determine your overall quality of life. Take into consideration your benefits, such as medical and dental insurance and retirement plans, when thinking about choosing your next job.

RETIREMENT

Retirement planning is hard to think about regardless of your age. However, even if retirement seems to be years away, setting aside money or other similar assets for the purposes of collecting an income once you're past working age is crucial. Most people don't realize until it's too late that they don't have the means to live as they hoped with financial independence.

The process of putting together plans to retire involves two basic parts. The first part is to assess your lifestyle and determine the age you wish to retire. The second is to come up with a plan of action and to make the decisions necessary to get you closer to your goals. Remember, retirement planning is not something that can be done in a day, a week, or even a month. Speak with your financial adviser at the FSC to chart the best course for you and your family.

Thrift Savings Plan (TSP)

The Thrift Savings Plan (TSP) is a retirement savings and investment plan for federal employees and members of the uniformed services, including the Ready Reserve. Established by Congress in the Federal Employees' Retirement System Act of 1986, it offers the same types of savings and tax benefits that many private corporations offer their employees under 401(k) plans.

The TSP is a defined contribution plan, where the income you receive during your retirement depends directly on how much you and your agency (if they match your contributions) put into your account. Participation in TSP is optional, but it is a great start in working toward your future retirement. For more information on TSP, contributions, and withdrawals, see chapter 2.

Survivor Benefit Plan (SBP)

The Survivor Benefit Plan helps offset the loss of pay due to your death. It protects your spouse or other eligible family members by providing a monthly income in the event that your retirement pay is no longer available.

Like regular insurance, premiums are paid for SBP coverage, but only after you retire. Premiums are taken directly out of your retired pay, which is reduced so it doesn't count as income. This means less tax and less out-of-pocket cost for SBP. The basic SBP for a spouse pays a benefit equal to 55 percent of your retired pay. If you are in a position to consider the election of SBP, do so it carefully. Failure to accept or decline coverage may result in you being automatically covered at the full retirement pay amount.

Planning Ahead

It is never too early to begin planning for retirement. By planning ahead for retirement, you can significantly reduce anxiety about your future and increase the potential for having a successful retirement.

Planning for retirement begins with identifying the kind of retirement you want to have. Write down your goals and dreams for retirement, and then start taking action to make it happen. Periodically reviewing your retirement plan provides the opportunity to explore new options, make adjustments, and stay on track for realizing your dreams and goals.

Setting financial goals is a critical part of retirement planning because you will need to have sufficient financial resources to pay for your retirement. Knowing how much money will be needed after you stop working can be difficult to determine, as this is dependent on your lifestyle, circumstances, and spending habits.

Financial planning helps organize your retirement plan. Using a long-term approach to managing your finances enables you to achieve your goals and dreams. A professional, such as your banker, accountant, tax preparer, or certified financial planner, can help you develop a strategy to achieve your financial goals for retirement. This tactic entails careful planning, involving the following:

- *Savings and cash flow*—saving a percentage of your paycheck for retirement (usually 10 percent or more), analyzing spending and earning patterns, and taking steps to control debt
- *Investments*—using a diversified approach to investing in stocks, bonds, cash, and annuities that provides balance and minimizes risk
- *Insurance*—having adequate life, auto, liability, health, disability, renters/homeowners, and long-term care insurance

- *Taxes*—taking advantage of tax-deferred retirement savings accounts such as 401(k), 403(b), Individual Retirement Accounts (IRAs), Savings Incentive Match Plan for Employees (SIMPLE), Simplified Employee Pension (SEP), and other employer-sponsored retirement plans
- *Estate planning*—maintaining up-to-date documents such as a will, power of attorney (both general and medical), beneficiary designations, and a medical directive (also known as a living will)

You can make the transition to retirement much smoother by having a plan for what you want to do when you stop working. Staying active and involved in new interests and projects makes it easier to find a new sense of direction. The options are endless, so plan to enjoy a variety of activities such as travel, hobbies, volunteering, going back to school, sports and exercise.

Retirement planning is a process that evolves over time as you determine what is best for you. Creating a retirement plan and staying focused on achieving your goals and dreams enables you to minimize anxiety and maximize security as you approach retirement.

Chapter Two

Bank Accounts, Savings, and Investments

Even though interest rates are very low, it is still a good idea to have a bank or credit union account. One of the most important reasons is that almost all bank and credit union accounts are protected by the Federal Deposit Insurance Corporation (FDIC), a federal agency that ensures that customers' money is protected even if the bank fails. Keeping money under the mattress may give you quick access to it, but it is no good if your house burns down or you are robbed!

Checking accounts make bill paying very convenient, and you can even make check deposits without having to leave home using free apps supplied by the banks for smart phones. Your pay can be deposited directly into your bank account, and many accounts also provide you with credit and debit cards so that you can easily access your money. Many banks also offer incentives so that if you use your checking account to pay bills and make purchases, a sum is paid automatically into a savings account.

If you have an account with a bank or credit union, it will also likely be easy to get a loan or mortgage from that institution.

Many financial advisers suggest that you should have an amount saved equivalent to between three and six months of your outgoings in case you fall on hard times. While for many people that is a very difficult goal to achieve in these harsh economic times, it should not stop you from trying to save.

ELECTRONIC BANKING

For many consumers, electronic banking means twenty-four-hour access to cash through an automated teller machine (ATM) or direct deposit of paychecks into checking or savings accounts. But electronic banking now involves many different types of transactions.

Electronic banking, also known as electronic fund transfer (EFT), uses computer and electronic technology as a substitute for checks and other paper transactions. EFTs are initiated through devices like cards or codes that let you, or those you authorize, access your account. Many financial institutions use ATM or debit cards and personal identification numbers (PINs) for this purpose. Some use other forms of debit cards such as those that require, at the most, your signature or a scan. The federal Electronic Fund Transfer Act (EFT Act) covers some electronic consumer transactions.

Electronic Fund Transfers

EFT offers several services that consumers may find practical.

Automated teller machines, or twenty-four-hour tellers (ATMs), are electronic terminals that let you bank almost any time. To withdraw cash, make deposits, or transfer funds between accounts, you generally insert an ATM card and enter your PIN. Some financial institutions and ATM owners charge a fee, particularly to consumers who don't have accounts with them or on transactions at remote locations. Generally, ATMs must tell you they charge a fee and its amount on or at the terminal screen before you complete the transaction. Check the rules of your institution and ATMs you use to find out when or whether a fee is charged.

Direct deposit lets you authorize specific deposits, such as paychecks and Social Security checks, to your account on a regular basis. You also may preauthorize direct withdrawals so that recurring bills, such as insurance premiums, mortgages, and utility bills, are paid automatically. Be cautious before you preauthorize direct withdrawals to pay sellers or companies with whom you are unfamiliar; funds from your bank account could be withdrawn fraudulently.

Pay-by-phone systems let you call your financial institution with instructions to pay certain bills or to transfer funds between accounts. You must have an agreement with the institution to make such transfers.

Personal computer banking lets you handle many banking transactions via your personal computer. For instance, you may use your computer to view your account balance, request transfers between accounts, and pay bills electronically.

Debit card purchase transactions let you make purchases with a debit card, which also may be your ATM card. This could occur at a store or business, on the Internet or online, or by phone. The process is similar to using a credit card, with some important exceptions. While the process is fast and easy, a debit card purchase transfers money—fairly quickly—from your bank account to the company's account. So it's important that you have funds in your account to cover your purchase. This means you need to keep accurate records of the dates and amounts of your debit card purchases and ATM withdrawals in addition to any checks you write. Also be sure you know the store or business before you provide your debit card information, to avoid the possible loss of funds through fraud. Your liability for unauthorized use, and your rights for error resolution, may differ with a debit card.

Electronic check conversion converts a paper check into an electronic payment in a store or when a company receives your check in the mail. In a store, when you give your check to a cashier, the check is run through an electronic system that captures your banking information and the amount of the check. You're asked to sign a receipt and you get a copy for your records. When your check has been handed back to you, it should be voided or marked by the merchant so that it can't be used again. The merchant electronically sends information from the check (but not the check itself) to your bank or other financial institution, and the funds are transferred into the merchant's account. When you mail a check for payment to a merchant or other company, they may electronically send information from your check (but not the check itself) through the system, and the funds are transferred into their account. For a mailed check, you should still receive advance notice from a company that expects to send your check information through the system electronically. The merchant or other company might include the notice on your monthly statement or under its terms and conditions. The notice also should state whether the merchant or company will electronically collect from your account a fee—like a "bounced check" fee—if you have insufficient funds to cover the transaction.

Be especially careful in Internet and telephone transactions that may involve use of your bank account information, rather than a check. A legitimate merchant that lets you use your bank account information to make a purchase or pay on an account should post information about the process on their website or explain the process over the telephone. The merchant also should ask for your permission to electronically debit your bank account for the item you're purchasing or paying on. However, because Internet and telephone electronic debits don't occur face-to-face, you should be cautious about to whom you reveal your bank account information. Don't give this information to sellers with whom you have no prior experience or in situations when you have not initiated the call, or to companies that seem reluctant to provide information or discuss the process with you.

Not all electronic fund transfers are covered by the EFT Act. For example, some financial institutions and merchants issue cards with cash value stored electronically on the card itself. Examples include prepaid telephone cards, mass transit passes, and some gift cards. These "stored-value" cards, as well as transactions using them, may not be covered by the EFT Act. This means you may not be covered for the loss or misuse of the card. Ask your financial institution or merchant about any protections offered for these cards.

Disclosures

To understand your legal rights and responsibilities regarding your EFTs, read the documents you receive from the financial institution that issued your "access device" (that is, the card, code, or other means of accessing your account to initiate electronic fund transfers). Although the means vary by institution, it often involves a card and/or a PIN. No one should know your PIN except you and select employees of the financial institution. You also should read the documents you receive for your bank account, which may contain more information about EFTs.

Before you contract for EFT services or make your first electronic transfer, the institution must tell you the following information in a form you can keep:

A summary of your liability for unauthorized transfers

The telephone number and address of the person to be notified if you think an unauthorized transfer has been or may be made, a statement of the institution's "business days" (which is, generally, the days the institution is open to the public for normal business), and the number of days you have to report suspected unauthorized transfers

The type of transfers you can make, fees for transfers, and any limits on the frequency and dollar amount of transfers

A summary of your right to receive documentation of transfers, to stop payment on a preauthorized transfer, and the procedures to follow to stop payment

A notice describing the procedures you must follow to report an error on a receipt for an EFT or your periodic statement, or to request more information about a transfer listed on your statement, and how long you have to make your report

A summary of the institution's liability to you if it fails to make or stop certain transactions

Circumstances under which the institution will disclose information to third parties concerning your account

A notice that you may be charged a fee by ATMs where you don't have an account

In addition to these disclosures, you will receive two other types of information for most transactions: terminal receipts and periodic statements. Separate rules apply to passbook accounts from which preauthorized transfers are drawn. The best source of information about those rules is your contract with the financial institution for that account. You're entitled to a terminal receipt each time you initiate an electronic transfer, whether you use an ATM or make a point-of-sale electronic transfer. The receipt must show the amount and date of the transfer, and its type, such as "from savings to checking." When you make a point-of-sale transfer, you'll probably get your terminal receipt from the salesperson.

You won't get a terminal receipt for regularly occurring electronic payments that you've preauthorized, like insurance premiums, mortgages, or utility bills. Instead, these transfers will appear on your periodic statement. If the preauthorized payments vary, however, you should receive a notice of the amount that will be debited at least ten days before the debit takes place.

You're also entitled to a periodic statement for each statement cycle in which an electronic transfer is made. The statement must show the amount of any transfer, the date it was credited or debited to your account, the type of transfer and type of account(s) to or from which funds were transferred, and the address and telephone number for inquiries. You're entitled to a quarterly statement whether or not electronic transfers were made.

Keep and compare your EFT receipts with your periodic statements the same way you compare your credit card receipts with your monthly credit card statement. This will help you make the best use of your rights under federal law to dispute errors and avoid liability for unauthorized transfers.

Errors

You have sixty days from the date a periodic statement containing a problem or error was sent to you to notify your financial institution. The best way to protect yourself if an error occurs—including erroneous charges or withdrawals from an account, or for a lost or stolen ATM or debit card—is to notify the financial institution by certified letter, return receipt requested, so you can prove that the institution received your letter. Keep a copy of the letter for your records.

If you fail to notify the institution of the error within sixty days, you may have little recourse. Under federal law, the institution has no obligation to conduct an investigation if you've missed the sixty-day deadline.

Once you've notified the financial institution about an error on your statement, it has ten business days to investigate. The institution must tell you the results of its investigation within three business days after completing it and must correct an error within one business day after determining that the error has occurred. If the institution needs more time, it usually may take up to forty-five days, in most situations, to complete the investigation—but only if the money in dispute is returned to your account and you're notified promptly of the credit. At the end of the investigation, if no error has been found, the institution may take the money back if it sends you a written explanation.

An error also may occur in connection with a point-of-sale purchase with a debit card. For example, an oil company might give you a debit card that lets you pay for gas purchases directly from your bank account. Or you may have a debit card that can be used for various types of retail purchases. These purchases will appear on your periodic statement from the bank. In case of an error on your account, however, you should contact the card issuer (for example, an oil company or a bank) at the address or phone number provided by the company. Once you've notified the company about the error, it has ten business days to investigate and tell you the results. In this situation, it may take up to ninety days to complete an investigation, if the money in dispute is returned to your account and you're notified promptly of the credit. If no error is found at the end of the investigation, the institution may take back the money if it sends you a written explanation.

Lost or Stolen ATM or Debit Cards

If your credit card is lost or stolen, you can't lose more than $50. If someone uses your ATM or debit card without your permission, you can lose much more.

If you report an ATM or debit card missing to the card issuer before it's used without your permission, you can't be held responsible for any unauthorized withdrawals.

If unauthorized use occurs before you report it, the amount you can be held responsible for depends upon how quickly you report the loss to the card issuer.

If you report the loss within two business days after you realize your card is missing, you won't be responsible for more than $50 for unauthorized use.

If you fail to report the loss within two business days after you realize the card is missing, but do report its loss within sixty days after your statement is mailed to you, you could lose as much as $500 because of an unauthorized transfer.

If you fail to report an unauthorized transfer within sixty days after your statement is mailed to you, you risk unlimited loss. That means you could lose all the money in your account and the unused portion of your maximum line of credit established for overdrafts.

If you failed to notify the institution within the time periods allowed because of an extenuating circumstance, such as lengthy travel or illness, the issuer must reasonably extend the notification period. In addition, if state law or your contract imposes lower liability limits, those lower limits apply instead of the limits in the federal EFT Act.

Once you report the loss or theft of your ATM or debit card, you're no longer responsible for additional unauthorized transfers occurring after that time. Because these unauthorized transfers may appear on your statements, however, you should carefully review each statement you receive after you've reported the loss or theft. If the statement shows transfers that you did not make or that you need more information about, contact the institution immediately, using the special procedures provided for reporting errors.

Limited Stop-Payment Privileges

When you use an electronic fund transfer, the EFT Act does not give you the right to stop payment. If your purchase is defective or your order is not delivered, it's as if you paid cash. That is, it's up to you to resolve the problem with the seller and get your money back.

There is one situation, however, in which you can stop payment. If you've arranged for regular payments out of your account to third parties, such as insurance companies, you can stop payment if you notify your institution at least three business days before the scheduled transfer. The notice may be oral or written, but the institution may require a written follow-up within fourteen days of the oral notice. If you fail to provide the written follow-up, the institution's responsibility to stop payment ends.

Although federal law provides only limited rights to stop payment, individual financial institutions may offer more rights or state laws may require them. If this feature is important to you, you may want to shop around to be sure you're getting the best "stop-payment" terms available.

Other Rights

The EFT Act protects your right of choice in two specific situations regarding use of electronic fund transfers. First, the act prohibits financial institutions from requiring you to repay a loan by electronic transfer. Second, if you're required to receive your salary or government benefit check by EFT, you have the right to choose your institution.

Suggestions

If you decide to use EFT, keep these tips in mind:

Take care of your ATM or debit card. Know where it is at all times; if you lose it, report it as soon as possible.

Choose a PIN for your ATM or debit card that's different from your address, telephone number, Social Security number, or birthdate. This will make it more difficult for a thief to use your card.

Keep and compare your receipts for all types of EFT transactions with your periodic statements. That way, you can find errors or unauthorized transfers and report them.

Make sure you know and trust a merchant or other company before you share any bank account information or preauthorize debits to your account.

Be aware that some merchants or companies may use electronic processing of your check information when you provide a check for payment.

Review your monthly statements promptly and carefully. Contact your bank or other financial institution immediately if you find unauthorized transactions and errors.

Where to File Complaints

If you think a financial institution or company has failed to fulfill its responsibilities to you under the EFT Act, speak up. In addition, you may wish to complain to the federal agency listed below that has enforcement jurisdiction over that company.

State Member Banks of the Federal Reserve System
Consumer and Community Affairs
Board of Governors of the Federal Reserve System
20th & C Streets, NW, Mail Stop 801
Washington, DC 20551
www.federalreserve.gov

National Banks
Office of the Comptroller of the Currency
Compliance Management
Mail Stop 7-5
Washington, DC 20219
www.occ.treas.gov

Federal Credit Unions
National Credit Union Administration
1775 Duke Street
Alexandria, VA 22314
www.ncua.gov

Non-Member Federally Insured Banks
Office of Consumer Programs
Federal Deposit Insurance Corporation
550 17th Street, NW
Washington, DC 20429
www.fdic.gov

Federally Insured Savings and Loans, and Federally Chartered State Banks
Consumer Affairs Program
Office of Thrift Supervision

1700 G Street, NW
Washington, DC 20552
www.ots.treas.gov

Other Credit and Debit or ATM Card Issuers

The FTC works for the consumer to prevent fraudulent, deceptive, and unfair business practices in the marketplace and to provide information to help consumers spot, stop, and avoid them. To file a complaint or to get free information on consumer issues, visit http://www.ftc.gov or call toll-free 1-877-FTC-HELP (1-877-382-4357); TTY: 1-866-653-4261. The FTC enters Internet, telemarketing, identity theft, and other fraud-related complaints into Consumer Sentinel, a secure online database available to hundreds of civil and criminal law enforcement agencies in the United States and abroad.

WORKING WITH FINANCIAL AND INVESTMENT CONSULTANTS

There are lots of resources available at no charge that can help you begin your journey to financial independence. Choosing the right financial and investment consultant can be a difficult task, and disastrous if you choose wrongly. However, members of the military have a number of internal options that can be taken advantage of without seeking outside help.

In-person financial counseling is available in most locations through Military OneSource, in partnership with National Foundation for Credit Counseling (NFCC).

Military OneSource arranges for you to meet face-to-face with a financial consultant in your community. Up to twelve counseling sessions per issue, per calendar year are allowed for each eligible client. The sessions are available to military and family members located in the continental United States. For those who are unable to attend in-person counseling or in locations where it is not available, MOS will provide telephone consultations.

You Have a Personal Financial Management Team

Start on your installation by talking with your Personal Financial Management Program (PFMP) office, located in your military and family support center. These offices are present on all DoD military installations. Find location and contact information by going to MilitaryINSTALLATIONS and choosing "Personal Financial Management Services" under Program/Service. National Guard and Reserve personnel not located near a military installation can access information and personalized financial counseling assistance through Military OneSource by calling 1-800-342-9647 or by visiting

Military OneSource online. There will be no charge to you for using this information.

You Can Get a 10 Percent Guaranteed Rate of Return on Savings While You're Deployed

Military personnel have the opportunity to earn 10 percent interest on up to $10,000 in savings annually while deployed to or in support of a combat zone. Uniformed members of the armed forces can contribute to the Savings Deposit Program, which is administered by the Defense Finance and Accounting Service, after thirty consecutive days of deployment outside the United States, for as long as you are receiving hostile fire pay. Any military finance office in theater can help you establish an account and assist you in setting up the deposit method most convenient for you.

You Are Eligible for the Federal Thrift Savings Plan

The Thrift Savings Plan is a retirement savings and investment plan for federal employees and members of the uniformed services, including the Ready Reserve. If you contribute as little as $20 per payday, your savings could really stack up.

Here is an example to give you an idea of how much you could have: assume twenty years of service, a contribution of $40 per month, and an annual expected rate of return of 7 percent (rates of return will vary over time, of course). Your contributions would total $9,600, and your return would total $11,359. Altogether, after twenty years you would have $20,959.

Use the http://www.TSP.gov calculator to see just how much money you could make.

SAVINGS AND INVESTMENTS

Budgeting and debt reduction are usually the first topics of conversation when discussing finances, but now is a good time to discuss savings and investments. While this book will offer you some basic concepts in saving and investing, it is always best to talk over your personal financial situation with an expert financial planner.

Banks and Credit Unions

These two organizations offer savings accounts designed to help you reach both your short-and long-term financial goals. When opening a savings account you need to look for an FDIC-insured savings account that offers competitive interest rates on low account balances. Don't forget to check withdrawal and deposit guidelines, as some institutions limit the amount of free transactions available. While savings accounts do not provide a lot of interest earnings on what you put into an account, they are designed so that funds are easily accessible.

Investments

For higher-interest earnings, you then need to invest your money into a variety of institutions that can work for you in making you money. There are four common types of investments: stocks, bonds, mutual funds, and cash equivalents.

Stocks

When you purchase stocks or equities, you essentially become part owner of the company. Any profits that are allocated to the owners are referred to as dividends.

Unlike bonds, which provide a steady stream of income, stocks can be volatile and fluctuate in value on a daily basis. Many stocks don't pay dividends—in which case, the only way that you can make money is if the stock increases in value. This is something to consider when determining if stocks are the investment route you want to take. Keep in mind that when you buy a stock, you aren't guaranteed anything.

Bonds

The term bond commonly refers to any securities that are founded on debt. By purchasing a bond, you are lending out your money to a company or government. In return, they agree to give you interest on your money and to eventually pay you back the amount you lent them.

The main attraction of bonds is their relative safety. As investments, savings bonds are safe and stand by their promise of providing fixed interest rates. The downside is that because there is little risk involved, there is little potential return. As a result, the rate of return on bonds is generally lower than on other investments.

Mutual Funds

A mutual fund is a large collection of stocks and bonds. Mutual funds allow you to pool your money with a number of other investors. This enables the group to pay a professional manager to select specific securities for you. The advantage of mutual funds is the possibility of diversifying your financial investment over a large pool of investments.

The main advantage of a mutual fund is that you can invest your money without the time or the experience that are often needed to choose a sound investment. Unfortunately, it is not easy to predict the risk and rate of return through mutual funds, and depending on the professional expertise of mutual fund managers, your success is likely to vary.

Cash Equivalents

Cash equivalents are short-term investments that are typically very liquid and easily convert into cash. They include Treasury bills (T-bills), money market funds, and some short-term certificates of deposit (CDs). These investments are relatively safe and will not generally create huge returns on your investment. You typically receive about a 2–5 percent yield on your money, but the more you invest, the more you earn.

There are many different ways that you can best strategize your savings and investment opportunities by speaking to a financial planner.

Savings Resources

Choose to Save
http://choosetosave.org/

CNN Clark Howard Money Coach
http://www.cnn.com/CNN/Programs/clark.howard/

Financial health goals and fundamentals
http://www.msmoney.com/mm/financial_health/finhealth_index.htm

Military Saves
http://www.militarysaves.org/

An array of financial subjects important to women
http://www.BlueSuitMom.com/money

Features on a variety of financial matters such as savings, budgeting, and planning
http://www.wife.org/money_invest.htm

Save and Invest
http://www.saveandinvest.org/MilitaryCenter/index.htm

Savings Bonds Program (1-800-722-2678)
http://www.savingsbonds.gov/

Women and Retirement Savings
http://www.dol.gov/ebsa/publications/women.html

THRIFT SAVINGS PLAN

The Thrift Savings Plan (TSP) is a retirement savings and investment plan for federal employees and members of the uniformed services, including the Ready Reserve. It was established by Congress in the Federal Employees' Retirement System Act of 1986 and offers the same types of savings and tax benefits that many private corporations offer their employees under 401(k) plans.

The TSP is a defined contribution plan, meaning that the retirement income you receive from your TSP account will depend on how much you (and your agency, if you are eligible to receive agency contributions) put into your account during your working years and the earnings accumulated over that time.

TSP Regulations

The TSP regulations can be found in title 5 of the Code of Federal Regulations, Parts 1600–1690, and are periodically supplemented and amended in the Federal Register.

The purpose of the TSP is to give you the ability to participate in a long-term retirement savings and investment plan. Saving for your retirement through the TSP provides many advantages, including:

- automatic payroll deductions
- a diversified choice of investment options, including professionally designed lifecycle funds
- a choice of tax treatments for your contributions:

    1. traditional (pre-tax) contributions and tax-deferred investment earnings, and
    2. Roth (after-tax) contributions with tax-free earnings at retirement if you satisfy the IRS requirements

- low administrative and investment expenses
- agency contributions, if you are an employee covered by the Federal Employees' Retirement System (FERS)
- under certain circumstances, access to your money while you are still employed by the federal government

- a beneficiary participant account established for your spouse in the event of your death
- a variety of withdrawal options

If you are covered by FERS, the TSP is one part of a three-part retirement package that also includes your FERS basic annuity and Social Security. If you are covered by the Civil Service Retirement System (CSRS) or are a member of the uniformed services, the TSP is a supplement to your CSRS annuity or military retired pay.

TSP benefits differ depending on your retirement system (FERS, CSRS, or uniformed services). If you aren't sure which retirement system covers you, check with your personnel or benefits office.

Regardless of your retirement system, participating in the TSP can significantly increase your retirement income, but starting early is important. Contributing early gives the money in your account more time to increase in value through the compounding of earnings.

If you're a CSRS employee or a member of the uniformed services, you have to make a TSP contribution election through your agency or service to establish a TSP account. You do not receive agency contributions. TSP contributions are payroll deductions. You have to make a "contribution election" through your agency or service to:

- start your contributions if you were not automatically enrolled
- increase or decrease your contributions if you were automatically enrolled
- change the amount of your employee contributions or their tax treatment (traditional or Roth)
- stop your contributions

First, ask your personnel or benefits office whether your agency or service handles TSP enrollments through paper TSP forms, or electronically through automated systems such as Employee Express, EBIS, myPay, LiteBlue, or the NFC PPS.

Next, tell your personnel or benefits office how much you want to contribute and the tax treatment of your contributions through the agency's or service's electronic system or by way of a TSP-1 form.

You can get copies of these forms from the TSP website (http://www.tsp.gov) or from your agency or service. Return completed forms to your agency or service, not to the TSP. Your agency needs your information to set up your payroll deductions. Your election should be effective no later than the first full pay period after your agency or service receives it.

Employee Contributions—for FERS, CSRS, and Uniformed Services

There are two types of employee contributions:

- regular
- catch-up (for participants fifty and older)

You have to contribute the maximum of regular contributions to be eligible to make catch-up contributions.

You can also choose between two tax treatments for your contributions:

- traditional (pre-tax)
- Roth (after-tax)

Regular employee contributions are payroll deductions that come out of your basic pay before taxes are withheld (traditional contributions) or after taxes have been withheld (Roth contributions). Each pay period, your agency or service will deduct your contribution from your pay in the amount you choose (or the automatic enrollment amount of 3 percent) and send your contribution to the TSP. Your agency or service will continue to do this until you make a new TSP election to change your contribution or stop it, or until you reach the Internal Revenue Code (IRC) contribution limit. How do you know if the correct amount is coming out of your pay? Check your earnings and leave statement to verify the amount.

Special conditions for uniformed service members: In addition to basic pay, you can also contribute from 1 to 100 percent of any incentive pay, special pay, or bonus pay—as long as you elect to contribute at least 1 percent from basic pay. Your total contributions from all types of pay must not exceed the applicable IRC contribution limit.

You can elect to contribute from incentive pay, special pay, or bonus pay, even if you are not currently receiving them. These contributions will be deducted when you receive any of these types of pay. If you are receiving tax-exempt pay (i.e., pay that is subject to the combat zone tax exclusion), your contributions from that pay will also be tax-exempt. (Earnings on tax-exempt contributions designated as traditional will be taxed at withdrawal. Earnings on tax-exempt contributions designated as Roth will be tax-free at withdrawal, provided you meet the requirements detailed later in this section.)

Catch-up contributions are payroll deductions that participants who are age fifty or older may be eligible to make in addition to regular employee contributions.

Catch-up contributions are voluntary and can be either traditional pre-tax or Roth after-tax. To be eligible to make catch-up contributions, you must already be contributing an amount that will reach the IRC elective deferral limit by the end of the year. In the year you turn fifty, you can begin making catch-up contributions at any time. Each pay period, your agency or service will make your contribution to the TSP from your pay in the amount you choose. Your catch-up contributions will stop automatically when you meet the IRC catch-up contribution limit or at the end of the calendar year, whichever comes first. Your catch-up contributions will not continue from year to year; you have to make a new election for each calendar year.

Special conditions for uniformed service members: You can't make catch-up contributions from incentive pay, special pay, or bonus pay. What's more, your traditional catch-up contributions will stop if you are receiving tax-exempt pay in a combat zone. Only Roth catch-up contributions are allowed from tax-exempt pay.

Currently, members of the uniformed services do not receive matching contributions. However, the secretary of each individual service is allowed by law to designate particular critical specialties as eligible for matching contributions under certain circumstances.

How Much You Can Contribute

The Internal Revenue Code (IRC) places a number of specific limits on the dollar amount of contributions you can make to the TSP. These limits can change annually and are generally referred to as the "IRS limits" because the Internal Revenue Service (IRS) is responsible for calculating them each year. When the annual limits become available, the TSP announces them on the TSP website and the ThriftLine, as well as through its various publications.

The IRC *elective deferral limit* is the maximum amount of employee contributions that can be contributed in a calendar year. The elective deferral limit applies to the combined total of your tax-deferred traditional contributions and Roth contributions. The IRC elective deferral limit for 2015 is $18,000.

For members of the uniformed services, elective deferrals include all traditional and Roth contributions from taxable basic pay, incentive pay, special pay, and bonus pay. However, the elective deferral limit of $18,000 does *not* apply to traditional contributions made from tax-exempt pay earned in a combat zone. If you are a member of the uniformed services who is contributing to both a uniformed services and a civilian TSP account as a FERS employee, the elective deferral limit applies to the total amount of tax-deferred traditional employee and Roth contributions you make in a calendar year.

The IRC section 415(c) limit is an additional limit that the IRC imposes on the total amount of all contributions made on behalf of an employee to an eligible retirement plan in a calendar year. "All contributions" include employee contributions (tax-deferred, after-tax, and tax-exempt), Agency Automatic (1 percent) Contributions, and Agency Matching Contributions. For 2013, the section 415(c) limit is $51,000.

Members of the uniformed services should pay particular attention to this section 415(c) limit if they contribute from pay that is subject to the combat zone tax exclusion, because section 415(c) allows their contributions to exceed the elective deferral limit.

The IRC *catch-up contribution limit* is the maximum amount of catch-up contributions that can be contributed in a calendar year by participants age fifty and older.

It is separate from both the elective deferral limit imposed on regular employee contributions and the IRC section 415(c) limit imposed on employee contributions (tax-deferred, after-tax, and tax-exempt), Agency Automatic (1 percent) Contributions, and Agency Matching Contributions. For 2013, the limit for catch-up contributions is $5,500 under IRC section 414(v).

The TSP offers you two tax treatments for your employee contributions when you make a contribution election:

1. *Traditional TSP*. If you make traditional contributions, you defer paying taxes on your contributions and their earnings until you withdraw them. If you are a uniformed services member making tax-exempt contributions, your contributions will be tax-free; only your earnings will be subject to tax at withdrawal.
2. *Roth TSP*. If you make Roth contributions, you pay taxes on your contributions as you are making them (unless you are making tax-exempt contributions from combat pay) and get your earnings tax-free at withdrawal, as long as you meet the requirements to qualify.

Tax Liability

When you withdraw your money from the TSP, you will owe taxes on any traditional contributions (except contributions made from tax-exempt pay) and the earnings they have accrued. You can continue to defer these taxes by transferring or rolling over your TSP withdrawal payment to a traditional individual retirement account (IRA) or an eligible employer plan. You can also transfer or roll over your traditional funds to a Roth IRA, but you will have to pay taxes on the full amount in the year of the transfer.

If you have Roth contributions in your account, you have already paid taxes on them. You will not owe any further taxes on your Roth contributions, and you will not owe taxes on their earnings if your withdrawal payment is a "qualified distribution." In other words, if five years have passed since January 1 of the calendar year when you made your first Roth contribution *and* you have reached age 59½ or have a permanent disability, the entire

Roth portion of your account will be paid out tax-free. If your earnings are not qualified, you can defer paying taxes on them by transferring your payment to a Roth IRA or Roth account maintained by an eligible employer plan.

Retirement age and the penalty tax If you receive a TSP withdrawal payment before you reach age 59½, you may have to pay a 10 percent early withdrawal penalty tax on any taxable part of the distribution not transferred or rolled over. This penalty tax is in addition to the regular income tax you owe, but there are exceptions. In general, if you leave federal service in the year you turn age fifty-five or older, the 10 percent penalty tax does not apply to any withdrawal you make that year or later.

In addition, disability retirement approved by the Office of Personnel Management may not exempt you from the early withdrawal penalty tax. The IRS requirement is more stringent, and you will have to substantiate your claim of exemption with the IRS. There are other exceptions to the early withdrawal penalty tax. See the tax notice "Important Tax Information about Payments from Your TSP Account," which is available from the TSP website, your agency or service, or the TSP. The tax rules that apply to distributions from the TSP are complex, and you may also want to consult with a tax adviser or the IRS before you make any withdrawal decisions.

Retirement savings contributions credit You may be able to take a tax credit of up to $1,000 (up to $2,000 if filing jointly) for your TSP contributions. Eligibility depends on the amount of your modified adjusted gross income (AGI). For tax year 2015, your AGI must be no more than $61,000 if married filing jointly, $45,750 if head of household, or $30,500 if single, married filing separately, or qualifying widow(er). (These amounts are adjusted each year for inflation. For more information, see your tax adviser or refer to IRS Form 8880.)

The TSP will accept into the traditional balance of your TSP account:

- both transfers and rollovers of tax-deferred money from traditional individual retirement accounts (IRAs), SIMPLE IRAs, and eligible employer plans.

The TSP will accept into the Roth balance of your TSP account:

- transfers of qualified and nonqualified Roth distributions from Roth 401(k)s, Roth 403(b)s, and Roth 457(b)s.

If you don't already have a Roth balance in your TSP account, the transfer will create one.

The TSP will *not* accept into your Roth balance:

- rollovers of Roth distributions that have already been paid to you
- transfers or rollovers from Roth IRAs

Why Transfer?

Transferring money into your TSP account allows you to consolidate your retirement savings in one place. This makes it easier to evaluate whether you are on target to reach your retirement savings goals, and to make sure the right asset allocation to meet these goals is applied to all your savings.

Also, because of the TSP's now legendary low costs, your savings can grow faster. This is why record numbers of TSP participants have been moving money into the TSP over the years.

Investing in the TSP

The TSP offers you two approaches to investing your money:

- *The L Funds*. These are "lifecycle" funds that are invested according to a professionally designed mix of stocks, bonds, and government securities. You select your L Fund based on your "time horizon," the future date at which you plan to start withdrawing your money. Depending upon your plans, this may be as soon as you leave or further in the future.
- *Individual Funds*. You make your own decisions about your investment mix by choosing from any or all of the individual TSP investment funds (G, F, C, S, and I Funds).

These investment options are designed so you can choose either the L Fund that is appropriate for your time horizon, or a combination of the individual TSP funds that will support your personal investment strategy. However, you may invest in any fund or combination of funds. Because the L Funds are already made up of the five individual funds, you will duplicate your investments if you invest simultaneously in an L Fund and the individual TSP funds.

The L Funds The L Funds are designed for participants who may not have the time, experience, or interest to manage their TSP retirement savings. The five L Funds are:

L 2050—For participants who will need their money in the year 2045 or later.

L 2040—For participants who will need their money between 2035 and 2044.

L 2030—For participants who will need their money between 2025 and 2034.

L 2020—For participants who will need their money between 2015 and 2024.

L Income—For participants who are already withdrawing their accounts in monthly payments, or who plan to need their money between now and 2015.

For the most up-to-date L Fund asset allocations, visit the TSP website (www.tsp.gov), Investment Funds section, and choose "Fund Options."

The Individual Funds The TSP has five individual investment funds:

- The Government Securities Investment (G) Fund—The G Fund is invested in short-term U.S. Treasury securities. It gives you the opportunity to earn rates of interest similar to those of long-term government securities with no risk of loss of principal. Payment of principal and interest is guaranteed by the U.S. government. The interest paid by the G Fund securities is calculated monthly based on the market yields of all U.S. Treasury securities with more than four years to maturity; the interest rate changes monthly.
- The Fixed Income Index Investment (F) Fund—The F Fund is invested in a bond index fund that tracks the Barclays Capital U.S. Aggregate Bond Index. This is a broad index representing the U.S. government, mortgage-backed, corporate, and foreign government sectors of the U.S. bond market. This fund offers you the opportunity to earn rates of return that exceed money market fund rates over the long term (particularly during periods of declining interest rates).
- The Common Stock Index Investment (C) Fund—The C Fund is invested in a stock index fund that tracks the Standard & Poor's 500 (S&P 500) Stock Index. This is a market index made up of the stocks of five hundred large to medium-sized U.S. companies. It offers you the potential to earn the higher investment returns associated with equity investments.
- The Small Capitalization Stock Index (S) Fund—The S Fund is invested in a stock index fund that tracks the Dow Jones U.S. Completion Total Stock Market (TSM) Index. This is a market index of small and medium-sized U.S. companies that are not included in the S&P 500 index. It offers you the opportunity to earn potentially higher investment returns that are associated with "small cap" investments, but with greater volatility.
- International Stock Index Investment (I) Fund—The I Fund is invested in a stock index fund that tracks the Morgan Stanley Capital International EAFE (Europe, Australasia, Far East) Index. This is a broad international market index, made up primarily of large companies in twenty-two developed countries. It gives you the opportunity to invest in international stock markets and to gain a global equity exposure in your portfolio.

For a chart comparing these five funds that provides more information about each, see www.tsp.gov/investmentfunds/fundsoverview/comparisonMatrix.shtml.

Because the TSP funds are trust funds that are regulated by the Office of the Comptroller of the Currency and not by the Securities and Exchange Commission (SEC), they do not have ticker symbols (i.e., the unique identifiers assigned to securities [including mutual funds] registered with the SEC). You can, however, obtain additional information about the underlying indexes that certain TSP funds track by visiting the following websites:

- F Fund Barclays Capital U.S. Aggregate Bond Index (http://www.barcap.com)
- C Fund Standard & Poor's 500 Stock Index (http://www.standardandpoors.com)
- S Fund Dow Jones U.S. Completion Total Stock Market (TSM) Index (http://www.djindexes.com)
- I Fund Morgan Stanley Capital International EAFE Stock Index (http://www.msci.com)

Fund Risks There are various types of risk associated with the TSP funds. There is no risk of investment loss in the G Fund. However, investment losses can occur in the F, C, S, and I Funds. Because the L Funds are invested in the individual TSP funds, they are also subject to the risks to which those underlying funds are exposed. These risks include:

- Credit risk—The risk that a borrower will default on a scheduled payment of principal and/or interest. This risk is present in the F Fund.
- Currency risk—The risk that the value of a currency will rise or fall relative to the value of others in the I Fund because of fluctuations in the value of the U.S. dollar in relation to the currencies of the twenty-two countries in the EAFE index.
- Inflation risk—The risk that your investments will not grow enough to offset the effects of inflation. This risk is present in all five funds.
- Market risk—The risk of a decline in the market value of the stocks or bonds. This risk is present in the F, C, S, and I Funds.
- Prepayment risk—A risk associated with the mortgage-backed securities in the F Fund. During periods of declining interest rates, homeowners may refinance their high-rate mortgages and prepay the principal. The F Fund must reinvest the cash from these prepayments in current bonds with lower interest rates, which lowers the return of the fund.

Choosing Your Own Investment Mix If you decide not to invest in the L Funds and you would rather choose your own investment mix from the G,

F, C, S, and I Funds, remember that your investment allocation is one of the most important factors affecting the growth of your account. If you prefer this approach, keep the following points in mind:

Consider both risk and return. Over a long period of time, the F Fund (bonds) and the C, S, and I Funds (stocks) have higher potential returns than the G Fund (government securities). But stocks and bonds also carry the risk of investment losses, which the G Fund does not. However, investing entirely in the G Fund may not give you the returns you need to keep up with inflation or meet your financial needs.

You need to be comfortable with the amount of risk you expect to take. Your investment comfort zone should allow you to use a long-term strategy so that you are not chasing market returns during upswings or abandoning your investment strategy during downswings.

You can reduce your overall risk by diversifying your investments. The five individual TSP funds offer a broad range of investment options, including government securities, bonds, and domestic and foreign stocks. Generally, it's best not to put "all of your eggs in one basket."

The amount of risk you can sustain depends upon your investment time horizon. The more time you have before you need to withdraw your account, the more risk you may be able to take. (This is because early losses can be offset by later gains.)

Periodically review your investment choices. Check the distribution of your account balance among the funds to make sure that the mix you chose is still appropriate for your situation. If not, rebalance your account to get the allocation you want. You can rebalance your account by making an interfund transfer.

Contribution Allocations and Interfund Transfers There are two types of investment transactions you can make:

Contribution allocations. A contribution allocation specifies how you want to invest new money *going into* your TSP account. Your contribution allocation will apply to all future deposits to your account. These include employee contributions; agency contributions (if you are FERS); any special pay, incentive pay, or bonus pay that you contribute as a member of the uniformed services; any money you move into the TSP from other retirement plans; and any TSP loan payments. Your contribution allocation *will not* affect money that is already in your account. Your contribution allocation will remain in effect until you submit a new one.

Interfund transfers. An interfund transfer moves the money *already in* your account among the TSP investment funds. When you make an interfund transfer, you choose the new percentage you want invested in each fund. You cannot move specific dollar amounts among the funds. Also, you cannot move specific types of money among the funds. For example, if you have traditional (including tax-exempt) and Roth balances in your account, your interfund transfer will move a proportional amount from each type of money into the funds that you have specified.

Interfund transfers are not unlimited. Each calendar month, your first *two* interfund transfers *may* be used to redistribute money in your account among any or all of the TSP funds. After the first two, your interfund transfers can *only* move money into the Government Securities Investment (G) Fund (in which case, you will increase the percentage of your account held in the G Fund by reducing the percentage held in one or more of the other TSP funds). If you have both a civilian and a uniformed services account, the following rules apply to each account separately.

Making a contribution allocation or interfund transfer. You can make either of these transactions on the TSP website or the ThriftLine (using the automated system, or by speaking to a TSP participant service representative). To make a contribution allocation or interfund transfer on the website, you will need your TSP account number (or customized user ID) and your Web password. To make a contribution allocation or interfund transfer on the ThriftLine, you will need your account number and your four-digit ThriftLine PIN (or press 3 to speak to a participant service representative). Contribution allocations or interfund transfers made on the TSP website or the ThriftLine before 12 noon Eastern time are generally processed that business day. You will receive a confirmation of your transaction in the mail or by e-mail, if you used the website for your transaction and chose that option.

Administrative Expenses

TSP expenses (i.e., the cost of administering the program) include management fees for each investment fund and the costs of operating and maintaining the TSP's recordkeeping system, providing participant services, and printing and mailing notices, statements, and publications. TSP expenses are lower than the industry average.

These expenses are paid primarily from the forfeitures of Agency Automatic (1 percent) Contributions of FERS employees who leave federal service before they are vested, other forfeitures, loan fees, and—because those forfeitures and fees are not sufficient to cover all of the TSP's expenses—earnings on participants' accounts.

The effect of administrative expenses (after forfeitures) on the earnings of the G, F, C, S, and I Funds is expressed as an expense ratio for each fund. The expense ratio for a fund is comprised of the total administrative expenses charged to that fund during a specific period, divided by that fund's average balance for that period.

Since the L Funds do not have any unique administrative expenses, the L Funds do not have any additional charges. Therefore, the L Fund administrative expense ratios are weighted averages of the expense ratios of the G, F, C, S, and I Funds.

Your share of TSP net administrative expenses is based on the size of your account balance. For example, the G Fund's expense ratio for 2014 was .029 percent. Therefore, if you invested in the G Fund in 2014, earnings were reduced by 29¢ per $1,000 of your G Fund balance.

TSP Loans, Withdrawals, and Refunds

Because the purpose of the TSP is for you to save money for your retirement, there are rules that restrict when and how you may take money out of your account while you are still employed. Once you leave federal service, however, you can take your money out at any time. However, the IRS may impose an early withdrawal penalty tax on the disbursement, depending on your employment status, when you take the disbursement, and how you receive the funds.

There are three ways to get your money out of the TSP:

- A *loan*
- An *in-service withdrawal* (i.e., a withdrawal while you are still employed by the federal government)
- A *post-separation withdrawal* (i.e., a withdrawal after you separate from service)

Any loan or withdrawal you take from your account will be paid proportionally from your traditional and Roth balances, and from each TSP fund in which you have investments. (The same is true for tax-exempt contributions in your traditional and Roth balances if you are a member of the uniformed services.) For example, you cannot request a loan or withdrawal from only the taxable portion of your traditional balance that is invested in the G Fund. If you have both traditional and Roth balances and you are invested in five TSP funds, both balances and all your fund investments will be impacted by your loan or withdrawal.

Loans When you take a loan, you are borrowing your own contributions and the earnings on those contributions.

When your loan is approved, the amount of the loan is removed from your TSP account. As you repay your loan, your loan repayments restore the amount of your loan, plus the interest you pay to your account.

You repay your loan with interest. The interest rate is the interest rate for the G Fund at the time your loan application is processed. The TSP also charges a processing fee of $50 for each loan. This fee is used to cover the cost of processing and servicing your loan. It is *deducted from the amount* of the loan that you receive.

Before you take a loan, consider that your loan costs are not limited to the interest and fee that you pay. The cost of a loan can be much more far-reaching.

When you borrow from your account, you miss out on the earnings that might have accrued on the money you borrowed. Even though you must pay the money back to your account with interest, the interest you pay to your account may be less than what you might have earned if you had kept the money in the TSP. Further, if you have an outstanding loan when you leave federal service, you must pay it back within ninety days or the outstanding balance will be treated as taxable income.

There are two types of TSP loans:

A general-purpose loan
A loan for the purchase or construction of a primary residence

You can have only one general-purpose and one residential loan outstanding at a time.

The total amount that you can borrow is limited to your own contributions and the earnings on those contributions. You cannot borrow less than $1,000 or more than $50,000. You can find out the amount you may be eligible to borrow from your TSP account by visiting the TSP website or calling the ThriftLine. You can also use the Estimate Loan Payments calculator on the TSP website to estimate your loan payment amount before you request a loan.

You do not need to provide any type of documentation for a general-purpose loan. However, you will need to provide documentation for a residential loan.

You must wait sixty days from the time you pay off one loan until you are eligible to request another loan of the same type.

Loan repayments are made through payroll deductions. They are deducted from your pay each pay period in the amount to which you agreed. If your agency or service does not deduct your loan payment from your pay, *you must submit the loan payment directly to the TSP with a TSP Loan Payment Coupon (Form TSP-26). You are responsible for your loan payments.*

You can also make additional payments or pay off your loan early by check or money order using the loan payment coupon, available at http://www.tsp.gov. And you can reamortize your loan to change the amount of your payment, number of payments, or repayment period.

You must repay your general-purpose loan within five years. Residential loans must be repaid within fifteen years.

If you fail to repay your loan in accordance with your loan agreement (or your most recent reamortization), or you do not repay your loan when you separate from service, the TSP will report a taxable distribution to the IRS.

You *will owe income taxes* on the taxable amount of the outstanding balance of the loan and possibly an early withdrawal penalty tax.

You *will not owe income taxes* on any part of your outstanding loan amount that consists of tax-exempt or Roth contributions. You *will owe taxes* on the earnings on tax-exempt contributions that were part of your traditional balance. The following conditions apply to Roth earnings:

- If the taxable distribution is declared because you separate from service, any *qualified Roth earnings will not be subject to tax*. Roth earnings that are not qualified will be subject to tax.
- If the taxable distribution is declared for another reason (such as a default on your loan), your *Roth earnings will be taxed, even if they were already qualified* (or eligible to be paid out tax-free).

Note: If you have two TSP accounts and you want to combine your accounts, you must close any loan in the account you are moving before the accounts can be combined.

If you are a married FERS or uniformed services participant, your spouse must consent to your loan by signing the loan agreement. If you are a married CSRS participant, your spouse will be notified of your loan. These rules apply even if you are separated from your spouse.

There are exceptions to these rights, but exceptions are rarely granted. See Form TSP-16, "Exception to Spousal Requirements" (U-16, uniformed services), for more information.

If you have a TSP loan, your payments must continue because, for bankruptcy purposes, a TSP loan is not a debt, and the TSP is not your creditor. Therefore, the bankruptcy court does not have jurisdiction over your TSP loan. For more information, see the TSP fact sheet "Bankruptcy Information." Different rules apply to bankruptcies filed since October 17, 2005.

For a detailed explanation of the TSP loan program, your obligations if you take a loan, and the consequences of not repaying a loan, read the TSP booklet *Loans*. For information about outstanding loans, you can check your earnings and leave statement, your participant statements, the TSP website, or the ThriftLine. You can also contact the TSP.

In-Service Withdrawals In-service withdrawals (i.e., withdrawals from your account while you are still employed) are available to all active participants. The TSP does not charge a fee for making an in-service withdrawal. However, the overall impact on your retirement savings may be significant.

You must pay federal, and in some cases, state income taxes on the taxable portion of the withdrawal, and you may also be subject to a 10 percent early withdrawal penalty tax. More importantly, if you make a financial hardship in-service withdrawal, the overall impact can be even greater because you cannot contribute to the TSP for six months following your withdrawal. If you are a FERS employee, that means you will also not receive any agency matching contributions during that time.

There are two types of in-service withdrawals:

- A *financial hardship* in-service withdrawal
- An *age-based* in-service withdrawal

Financial hardship in-service withdrawal. You can make a financial hardship in-service withdrawal if you can certify, under penalty of perjury, that you have a financial hardship as a result of a recurring negative cash flow, legal expenses for separation or divorce, medical expenses, or a personal casualty loss. You may withdraw only your contributions and any earnings those contributions have accrued. You can request $1,000 or more; however, the amount that you request cannot exceed the actual amount of your certified financial hardship. Further, you may not make contributions to your account (and if you are FERS, you will not receive the associated matching contributions) for six months after the disbursement of your funds.

You can make an age-based in-service withdrawal anytime after you reach age 59½, as long as you are still a civilian federal employee or a member of the uniformed services. You may withdraw part or all of your vested account balance. You can request a dollar amount of $1,000 or more, or your entire account balance (even if it is less than $1,000).

You are permitted to make only one age-based in-service withdrawal. If you make one, you will not be eligible to make a partial withdrawal from your account after you separate from service.

If you are a married FERS or uniformed services participant, your spouse must consent to your in-service withdrawal. If you are a married CSRS participant, the TSP must notify your spouse before an in-service withdrawal can be made. These rules apply even if you are separated from your spouse.

There are exceptions to these rights, but exceptions are rarely granted. For more information, see Form TSP-16 (or U-16 for members of the uniformed services), "Exception to Spousal Requirements."

You must pay federal income taxes on the taxable portion of in-service withdrawals when they are paid directly to you. You will owe taxes on the portion of your withdrawal that comes out of your traditional balance (excluding tax-exempt contributions). You can retain the tax-deferred status of the traditional portion of your age-based withdrawal by transferring it to a traditional IRA or eligible employer plan. (You can also transfer it to a Roth IRA, but you would have to pay taxes on the transfer in the year it is made.)

You will not pay federal income taxes on the portion of your in-service withdrawal that comes from your Roth contributions, and you will only pay taxes on the earnings if they are not qualified. However, you can transfer the Roth portion of your withdrawal to a Roth IRA or a Roth account maintained by an eligible employer plan. Financial hardship in-service withdrawals may be subject to an early withdrawal penalty tax if you are younger than age 59½ when you make your withdrawal. For more detailed information about the tax rules, see the TSP tax notice "Important Tax Information about Payments from Your TSP Account."

Withdrawals after You Separate If your vested account balance is *$200 or more* after you leave federal service, you can leave your money in the TSP until later, or you can withdraw all or a portion of your account. If you leave your money in the TSP after you separate from service, be sure to keep your address up-to-date so that the TSP can reach you.

Any withdrawal from your account will be made up of a proportional amount of traditional (non-Roth) and Roth money.

If your vested account balance is *less than $200* when you leave federal service, the TSP will automatically send you a check for the amount in your account. The check will be mailed to the address in your TSP account record.

You cannot leave this money in the TSP or make any other withdrawal election.

If you decide to leave money in the TSP after you separate from either the uniformed services or federal civilian service, you will be able to combine your TSP accounts by submitting Form TSP-65, "Request to Combine Uniformed Services and Civilian TSP Accounts." Money that you transfer will be deposited as employee contributions into the traditional or Roth balance of the combined account based on the way it was identified in the original account.

There are restrictions about how and when accounts can be combined. For example, you can only combine the money from the account related to your separation into your other account (and if you have a loan in the account you are moving, you must close it before you can combine your accounts). Also, tax-exempt contributions (i.e., contributions from combat zone pay) in your uniformed services TSP account may not be transferred to your civilian TSP account unless they are part of your Roth balance. Tax-exempt contributions that are part of your traditional (non-Roth) balance must remain in your uniformed services account.

There are two types of post-separation withdrawals:

A *partial withdrawal*
A *full withdrawal*

Partial withdrawal. You can take out $1,000 or more and leave the rest in your account until you decide to withdraw it at a later date. You may make only one partial withdrawal from your account. If you made an age-based in-service withdrawal, you are not eligible for a partial withdrawal.

Full withdrawal. You choose how your entire account will be distributed using one—or any combination—of three withdrawal options available to you:

A *single payment*
A series of *TSP monthly payments*
A *life annuity purchased for you by the TSP*

A single payment allows you to withdraw your entire TSP account at one time in one payment. It is sometimes referred to as a "lump sum."

TSP monthly payments allow you to withdraw your entire account in a series of payments that will be paid to you each month from your TSP account. You can ask for a specific dollar amount each month, or you can have the TSP calculate a monthly payment based on your life expectancy. If you choose a specific dollar amount, it must be at least $25.

At any time while you are receiving monthly payments, you can ask the TSP to stop the monthly payments and pay you your remaining account balance in a single payment. Also, once a year, you have the opportunity to make changes to the dollar amount of the monthly payments you are receiving. You also have the opportunity to make a one-time switch to receiving monthly payments based on a dollar amount rather than monthly payments based on life expectancy.

An annuity pays a benefit to you (or to your survivor) every month for life. The TSP purchases the annuity on your behalf from a private insurance company. You can have the TSP purchase an annuity with all or any portion of your account balance when you request a full withdrawal. In general, the amount you use for the purchase of an annuity must be $3,500 or more.

If you choose a life annuity and you have only one type of balance (traditional or Roth) in your TSP account, you must have at least $3,500 in your account at the time your annuity is purchased. If you are using only a portion of your account for an annuity, the percentage you choose when requesting your withdrawal must equal $3,500 or more of your vested account balance.

If you choose a life annuity and you have both a traditional balance and a Roth balance in your TSP account, the minimum threshold of $3,500 applies to *each balance separately*. You may choose to purchase an annuity as long as you have $3,500 in either your traditional or Roth balance. The TSP will purchase *two* of the same type of annuity (one with the traditional balance and one with the Roth balance). You cannot choose different annuities for each type of balance.

The following rules also apply:

If you choose to use 100 percent of your TSP account to purchase an annuity and both balances are below $3,500, your withdrawal form will be rejected. If you have both a traditional balance and a Roth balance and at least one of the balances is at least $3,500, the TSP will purchase an annuity with the balance that is at least $3,500 and pay the other balance directly to you as a cash payment.

Alternatively, *if you choose an annuity as part of a mixed withdrawal*, any amount(s) that cannot be used to purchase the requested annuity will be split proportionally and distributed according to the other withdrawal option(s) you have chosen.

You have a choice of three basic annuity types:

- A *single life annuity*—paid only to you during your lifetime.
- A *joint life annuity with your spouse* —paid to you while you and your spouse are alive. When one of you dies, payments are made to the survivor for the rest of his or her life.
- A *joint life annuity with someone (other than your spouse) who has an insurable interest in you*—paid to you while you and the person you choose are alive. When one of you dies, payments are made to the survivor for his or her life.

If you elect a joint annuity, you may be able to choose between a 50 percent or 100 percent payment option to the survivor. Some additional annuity features may also be available, depending on the basic annuity type you choose. You may be able to request "cash refund," "ten-year certain," or "increasing payment" features. The available annuities and their features are explained in detail in the booklet *Withdrawing Your TSP Account after Leaving Federal Service*.

If you are a married FERS or uniformed services participant, your spouse must consent to your partial withdrawal. If you are a married CSRS participant, the TSP must notify your spouse before a partial withdrawal can be made.

If your vested account balance at the time of your full withdrawal is more than $3,500, your withdrawal will be subject to federal law regarding spouses' rights. The following rules apply even if you are separated from your spouse:

If you are a married FERS or uniformed services participant, your spouse is entitled to an annuity with a 50 percent survivor benefit, level payments, and no cash refund feature. Your spouse must waive the right to this particular annuity unless you use your entire account balance to purchase it.

If you are a married CSRS participant, the TSP must notify your spouse before it can process your withdrawal, regardless of which withdrawal option you choose.

In the event of your death, your account will be distributed to the beneficiary or beneficiaries you designate on the TSP's Designation of Beneficiary form. If you do not designate beneficiaries to receive your account, it will be disbursed according to the following order of precedence required by law:

- To your spouse
- If none, to your child or children equally, and to descendants of deceased children by representation
- If none, to your parents equally or the surviving parent
- If none, to the appointed executor or administrator of your estate
- If none, to your next of kin who is entitled to your estate under the laws of the state in which you resided at the time of your death

For this order of precedence, a child includes a natural child or an adopted child but does not include a stepchild who has not been adopted. A parent does not include a stepparent, unless your stepparent has adopted you. "By representation" means that if your child predeceases you, his or her share will be divided equally among his or her children.

A will or any other document (such as a prenuptial agreement) is not valid for the disposition of your TSP account.

If you wish, you can designate a person or persons, your estate, or a trust to receive your TSP account after your death. To designate a beneficiary or beneficiaries, you *must* use Form TSP-3, "Designation of Beneficiary." *The completed form must be properly signed, witnessed, and received by the TSP on or before the date of your death.*

CREDIT CARDS

There are an estimated 550 million credit cards in circulation in the United States. Americans used their credit cards to spend an estimated $2 trillion in 2014, and credit card debt is estimated at $889 billion. The CARD Act, which was signed into law more than three years ago, made credit card costs more reliable—with less risk of unexpected rate increases or other charges.

But despite this progress, a recent study by J. D. Power found that roughly two-thirds of cardholders say they don't completely understand how their cards work. And, as indicated in a recent Consumer Financial Protection Bureau (CFPB) report on credit card complaints received by the Bureau, difficulty understanding the terms of their cards is a contributing factor in many consumer complaints.

Credit card agreements—contracts that consumers receive when they sign up—include information about the costs, features, and terms of the product. But while some companies have made improvements, agreements are often long, complicated, and written in legalese. Key information about interest rates, fees, billing, and payments is often surrounded by legal fine print.

A credit card's interest rate is the price you pay for borrowing money. For credit cards, the interest rates are typically stated as a yearly rate, called the annual percentage rate (APR).

A fixed-rate APR or fixed APR sets an APR that does not fluctuate with changes to an index. This does not mean that the interest rate will never change, but the issuer generally must notify you before the change occurs, and in most circumstances can apply the higher rate only to purchases and other transactions you make after you get the notice.

A variable-rate APR or variable APR changes with the index interest rate, such as the prime rate published in the *Wall Street Journal*. The cardholder agreement will say how a card's APR can change over time.

With most credit cards, you can avoid paying interest on purchases if you pay your balance in full each month. The period between the end of a billing cycle and the date your payment is due is referred to as a "grace period." Card issuers that provide a grace period must establish procedures to assure that bills are mailed or delivered at least twenty-one days before they are due.

If you do not pay your balance in full, you will generally be charged interest on the unpaid portion of the balance, and interest will be charged on purchases in the new billing cycle starting on the date each purchase is made.

Different card issuers have different rules for determining when they charge interest. In general, once a card issuer begins to charge interest, it will continue to do so until it receives your payment. This means that if you have been carrying a balance, you will be charged interest—sometimes called "residual interest"—from the time your bill was sent to you until the time your payment is received by your card issuer.

With credit cards, grace periods typically apply only to purchase transactions. If you use your card to get a cash advance or use a check you received from your card issuer, generally you will start paying interest as of the date of the transaction. Most card issuers also charge a different interest rate if you use your credit card to get cash or to write a check. Your statement generally must show the APR that applies to cash and cash-like transactions and the amount of the balance that falls in that category. There may be more than one such APR—for example, one that applies to cash advances and a different APR for checks.

Your cardholder agreement must include the rules your card issuer applies to determine which transactions fall into which categories and must list the different interest rates. You should be able to find a copy of the agreement on your card issuer's website, and you can get a copy from your card issuer. If you have any questions, you should contact your card issuer.

A daily periodic interest rate is calculated by dividing the annual percentage rate (APR) by either 360 or 365, depending on the card issuer. The resulting daily periodic interest rate is then used to calculate interest by multiplying the rate by the amount owed at the end of each day. This amount is then added to the previous day's balance, which means that interest is compounding on a daily basis.

Your Rights under the Servicemembers Civil Relief Act (SCRA)

Don't lose money on your credit cards! The following are some questions and answers for those who are

entering the military and have an existing credit card balance
on active duty and have credit card debt from before they were on active duty
in the National Guard/Reserves and have been called up to active duty

Q. I have an existing credit card balance. Can I get any relief from the finance charges?

You have rights. As a general matter, when you enter active duty, you should notify your card issuer. The maximum interest rate you can be charged on any amount you owed before entering active-duty service is 6 percent. For this purpose, interest includes not just periodic interest charges, but also other finance charges and fees related to the debt. One such fee is an annual fee.

For members of the full-time active-duty military, SCRA protections begin the day you enter military service. For a Reservist or Guardsman, SCRA protections begin the day you receive your mobilization orders.

To get the benefit of the SCRA, you must notify your credit card company of your active-duty status in writing. You must send a written letter to the card issuer and include a copy of your orders. Include in your letter a request to reduce your interest to 6 percent while you are on active duty.

Some credit card issuers may even be willing to reduce your interest rate further than the SCRA requires.

Q. I have been on active duty and have returned home. I just learned that I could have gotten a reduction in the interest rate on my credit card while I was on active duty. Is it too late for me to get this reduction?

You have up to 180 days after you are released from active duty to let a lender know that you were on active duty. You should write your credit card company and send a copy of your military orders. If you do so within this 180-day time period, you are entitled to have your interest rate reduced to 6 percent. This reduction is effective for the period from the date you entered active duty through the date you were released from active-duty status.

Q. If I ask my credit card company to reduce the interest rate on my balance while I am on active duty, can it close my account or reduce my credit line?

No. Under the law a lender cannot revoke or reduce your credit because you have exercised your right to a reduced interest rate under the Servicemembers Civil Relief Act.

Q. If I ask my credit card company to reduce the interest rate on my balance while I am on active duty, can that hurt my credit rating?

No. It is against the law for a lender to make an adverse credit report because you exercised your rights under the Servicemembers Civil Relief Act.

Q. Can I get any reduction in the interest rate I am charged on any new purchases I make using my credit card while I am on active duty?

The law regulates only the interest rate on amounts that you owed at the time you entered the military service or were called up to active duty. If you make purchases with your card while on active duty, you can be charged your regular APR on this new balance. However, your card issuer must apply toward the new balance any monthly payment you make that exceeds the minimum amount due. This will reduce the total amount of interest you pay on the card.

Some credit card companies may give you a reduced interest rate on new purchases as well as on the balance you owed when you went on active duty. You should check with your credit card company to see what, if anything, it will do for you.

Q. I believe that my rights as a service member have been violated by my credit card issuer. What should I do?

You should contact the nearest Armed Forces Legal Assistance Program office. Dependents of service members can also contact or visit local military legal assistance offices where they reside. You can get help online to find an office, whether you are within the continental United States or anywhere else worldwide. Another potential source of assistance that may be helpful even if you are no longer on active duty is the American Bar Association.

Credit Scores

Consumer reporting agencies (CRAs) are companies that gather, organize, standardize, and disseminate consumer information, especially credit information. Each of the nationwide CRAs—Equifax, TransUnion, and Experian—have their own proprietary generic scoring models to predict credit performance. These models were originally developed for use by lenders to predict performance on credit obligations, but they are now primarily sold as educational scores to consumers.

The Dodd-Frank Wall Street Reform and Consumer Protection Act directed the Consumer Financial Protection Bureau (CFPB) to compare credit scores sold to creditors and those sold to consumers by nationwide CRAs and determine whether differences between those scores disadvantage consumers. CFPB analyzed credit scores from 200,000 credit files from each of the three major nationwide CRAs.

Some of the proprietary generic scores sold by the CRAs are as follows:

Equifax: "Equifax Credit Score." Produces scores in the range 280–850.
Experian: "Experian Plus Score." Produces scores in the range 330–830.
TransUnion: "TransRisk New Account Score." Produces scores in the range 300–850.

While consumers can obtain free annual credit reports from the nationwide CRAs, they typically have to pay for credit scores.

Many lenders use specific score levels as thresholds to determine whether consumers will qualify for a particular loan or interest rate. For a majority of consumers, the scores produced by different scoring models provide similar information about the relative creditworthiness of the consumers. That is, if a consumer had a good score from one scoring model, the consumer likely had a good score on another model. For a substantial minority, however, different scoring models gave meaningfully different results.

Consumers Should Check Their Credit Reports for Accuracy and Dispute Any Errors

Credit scores are calculated based on information in a consumer's credit file. Regardless of the credit scoring model used, inaccurate adverse information in a consumer's file (e.g., unpaid accounts that are not the consumer's, accounts described as paid late that were paid on time) can hurt that consumer's credit score. Before shopping for major credit items, consumers should review their credit files for inaccuracies. Each of the nationwide CRAs is required by law to provide credit reports for free to consumers once every twelve months upon request. A consumer can obtain these reports at http://www.annualcreditreport.com/cra/index.jsp. Consumers can get information on this and the dispute process at http://www.consumerfinance.gov/askcfpb.

Consumers Should Shop for Credit

Regardless of variations in educational and commercial scores, or even among scoring models used by lenders (which were analyzed in this study in only a very limited and somewhat indirect manner), consumers benefit by shopping for credit. Even if provided the same score, lenders may offer different loan terms because they operate different risk models or face different competitive pressures. Consumers should not rule themselves out of seeking lower-priced credit due to assumptions about their credit score.

Some consumers are reluctant to shop for credit out of fear that they will harm their credit score. Many consumers are generally aware that inquiries by creditors can negatively impact their credit score. However, the potentially negative impact of inquiries on credit scores may be overblown.

For a substantial minority of consumers, the scores that consumers purchase from the nationwide CRAs depict consumers' creditworthiness differently from the scores sold to creditors. It is likely that, unaided, many consumers will not understand this fact or even understand that the score they have obtained is an educational score and not the score that a lender is likely to rely upon. Consumers obtaining educational scores may be confused about the usefulness of the score being sold if sellers of scores do not make it clear to consumers before the consumer purchases the educational score that it is not the score the lender is likely to use.

Credit Counseling

Living paycheck to paycheck? Worried about debt collectors? Can't seem to develop a workable budget, let alone save money for retirement? If this sounds familiar, you may want to consider the services of a credit counselor. Many credit-counseling organizations are nonprofit and work with you to solve your financial problems. But beware—just because an organization says it is "nonprofit" doesn't guarantee that its services are free or affordable, or that its services are legitimate. In fact, some credit counseling organizations charge high fees, some of which may be hidden, or urge consumers to make "voluntary" contributions that cause them to fall deeper into debt.

Most credit counselors offer services through local offices, the Internet, or on the telephone. If possible, find an organization that offers in-person counseling. Your financial institution, local consumer protection agency, and friends and family also may be good sources of information and referrals.

Choosing a Credit Counseling Organization

Reputable credit counseling organizations advise you on managing your money and debts, help you develop a budget, and usually offer free educational materials and workshops. Their counselors are certified and trained in the areas of consumer credit, money and debt management, and budgeting. Counselors discuss your entire financial situation with you, and help you develop a personalized plan to solve your money problems. An initial counseling session typically lasts an hour, with an offer of follow-up sessions.

A reputable credit counseling agency should send you free information about itself and the services it provides without requiring you to provide any details about your situation. If a firm doesn't do that, consider it a red flag and go elsewhere for help.

Once you've developed a list of potential counseling agencies, check them out with your state attorney general, your local consumer protection agency, and the Better Business Bureau. They can tell you if consumers have filed complaints about them. (But even if there are no complaints about them, it's not a guarantee that they're legitimate.)

The United States Trustee Program also keeps a list of credit counseling agencies that have been approved to provide prebankruptcy counseling. You can find a state-by-state list of government-approved organizations at http://www.usdoj.gov/ust. After you've done your background investigation, it's time for the most important research—you should interview the final "candidates."

Here are some questions to ask to help you find the best counselor for you:

- *What services do you offer?* Look for an organization that offers a range of services, including budget counseling and savings and debt management classes. Avoid organizations that push a debt management plan (DMP) as your only option before they spend a significant amount of time analyzing your financial situation.
- *Do you offer information?* Are educational materials available for free? Avoid organizations that charge for information.
- *In addition to helping me solve my immediate problem, will you help me develop a plan for avoiding problems in the future?*
- *What are your fees?* Are there set-up and/or monthly fees? Get a specific price quote in writing.
- *What if I can't afford to pay your fees or make contributions?* If an organization won't help you because you can't afford to pay, look elsewhere for help.
- *Will I have a formal written agreement or contract with you?* Don't sign anything without reading it first. Make sure all verbal promises are in writing.
- *Are you licensed to offer your services in my state?*
- *What are the qualifications of your counselors?* Are they accredited or certified by an outside organization? If so, by whom? If not, how are they trained? Try to use an organization whose counselors are trained by a nonaffiliated party.
- *What assurance do I have that information about me (including my address, phone number, and financial information) will be kept confidential and secure?*
- *How are your employees compensated?* Are they paid more if I sign up for certain services, if I pay a fee, or if I make a contribution to your organiza-

tion? If the answer is yes, consider it a red flag and go elsewhere for help. (Source: Federal Trade Commission)

Loans and Financial Aid

Loans

The dictionary defines loan as the act of lending something on condition of being returned; a grant of the temporary use of something. Many types of loans are available and each one has specific repayment conditions. You will be better prepared to secure a loan that is appropriate for your circumstances by having a basic understanding of loans.

The basic principle of a loan is that one person (the lender) agrees to loan something to another person (the borrower), and it is mutually understood that the borrower will return the item to the lender. Essentially, the borrower has incurred a debt to the lender. When dealing with monetary loans, the transaction must benefit both parties.

Most monetary loans are provided at a cost, known as interest on the debt, which provides an incentive for the lender to offer the loan. Monetary loans normally involve a legally binding, written contract specifying the conditions under which the money will be repaid. Individuals may consider numerous options for obtaining a loan such as financial institutions, banks, the government, friends, or family members. It is important to gather information from more than one lender to ensure you obtain a loan that is suitable for your situation.

Typically, there are three parts to a loan—the interest rate, term, and security:

Interest rate: The amount the lender charges you for the use of their money. The interest rate is usually a percentage of the loan amount and can be fixed (unchanging) or variable (subject to change).

Term: The term of a loan is the maximum amount of time the borrower is given to pay back the loan. Borrowers are usually permitted to repay the loan before the end of the term if desired. The normal rule of thumb is the longer the term, the higher the interest rate.

Security for the lender: All loans are either secured or unsecured.

A *secured loan* means you (the borrower) have guaranteed repayment of the loan by giving the lender the right to claim something you own (an asset). If the loan goes unpaid, the lender can take your asset to recover their money. Interest rates for secured loans are generally low because the lender has the security of knowing their investment will be returned.

With an *unsecured loan*, the borrower does not offer any assets. Interest rates are normally high for unsecured loans because the lender has no way of recouping their money if the borrower fails to repay the loan. For this reason, lenders often seek extra protection against losing their investment by having an additional person co-sign for the loan.

The type of loan you wish to obtain is commonly dependent on the reason you need to borrow money. For example:

- *Residential loans* (known as mortgages) provide funding for the purchase, refinance, or renovation of a residence.
- *Automobile loans* finance the purchase of automobiles.
- *Student loans* provide funding to attend school.
- *Lines of credit* are loans that allow a borrower to take and repay money as needed.
- *Debt consolidation loans* allow you to combine existing debts into one loan to lower your monthly repayments and reduce the total cost.
- *Home equity loans* allow you to release the value of your property as cash for almost any purpose.

It is important to carefully review the interest rate and terms of a loan before making a selection. Use caution when considering any loan that carries extremely high interest rates, as this may result in leading you deeper into debt and could end up costing a fortune.

The application process begins by completing forms with a loan representative. Some loans require the borrower to provide supporting documentation such as financial information and account statements. Lenders often charge application fees, which vary depending on the complexity of the loan.

Lenders generally look at the four C's when considering a loan for approval: credit, capacity, capital, and collateral.

1. *Credit* is your history of repayment. A poor credit history can eliminate you from consideration for loan approval.
2. *Capacity* is your ability to repay the loan. The lender reviews your income and debt to determine if you will be able to repay the loan.
3. *Capital* is your cash on hand. This is very important when a large sum of money is being requested. For example, when applying for a loan to purchase a home, you must show the lender that you have sufficient funds on hand to cover the down payment and closing costs.
4. *Collateral* is the security for the loan. Some loans require the borrower to provide property, cash, equipment, inventory, or other business assets to back the loan. The value of the collateral must equal or exceed the loan amount.

Once you have the loan money in hand, you must start to repay the lender according to the terms of the loan. Typically, monthly payments are made until the loan is repaid. If you are late in making a loan payment, a late fee may be charged. If you become seriously behind in your payments (e.g., more than ninety days late), your account may be turned over to a collection agency and the lender may take possession of the collateral.

Loans are great in that they allow you to borrow money for things that you really want or need. Finding the right loan, however, can be a challenge. Understanding the basics of loans, how they work, and careful planning can prepare you for success.

Financial Aid

VFW Veterans Foreign Wars Unmet Needs Program With the help of corporate sponsors, the VFW Foundation receives funding to establish, administer, and promote the Unmet Needs Program. Funds from donations are available to the five branches of service (Army, Navy, Air Force, Marines, and Coast Guard), as well as members of the Reserves and National Guard. Funds awarded by the program are offered in the form of grants—not loans—so recipients don't need to repay them. Eligibility criteria include the following:

- The service member has to have been active duty or discharged from active duty within thirty-six months prior to applying.
    - Can receive funds only once every eighteen months.
- The hardship must be primarily due to deployment or military service.
    - Hardships caused by civil, legal, or domestic misconduct are not eligible for the grant.
    - Hardships caused by financial mismanagement by self or others, or due to bankruptcy, are not eligible for the grant. Applicants with these situations will be provided with resource information and referrals to other agencies.
- The applicant must be the service member, or the applicant must be currently listed or eligible to be listed as a dependent of the service member under DEERS.

Persons eligible to apply on behalf of the military family in need include:

- Personnel
- Military unit point of contact—Family Assistance Center coordinator, commanding officer, medical hold case worker
- VA Representative or VFW Service Officer assisting with a VA claim

Expenses eligible for payment include:

- Housing expenses—mortgage, rent, repairs, insurance
- Vehicle expenses—payments, insurance, repairs
- Utilities, including the primary phone
- Food and personal items
- Children's clothing, diapers, formula, school or childcare expenses
- Medical bills, prescriptions and eyeglasses—the patient's portion for necessary or emergency medical care only
- Appliance repair

Ineligible expenses include:

- Credit cards, military charge/debit cards, retail store credit cards
- Personal, student, or payday loans
- Negative bank accounts
- Cable, Internet, secondary phone, or cell phone
- Cosmetic or investigational medical procedures and expenses
- Taxes—property or otherwise
- Child support or alimony
- IRS or military debt, or debt owed to a friend/family member
- Legal or educational expenses
- Furniture, electronic equipment, and vehicle rentals
- Down payments on homes or vehicles
- Reimbursements for items already paid for
- Bills obviously due to excessive use or personal mismanagement

Other Nonprofits There are many other nonprofit organizations offering similar support. These include the Navy USMC Relief Society, Operation Homefront, Injured Marine Semper Fi, Hope for the Warriors, Americasheroesatwork.gov, and Americasupportsyou.com. Your family support center can provide a full list of all such organizations available to you.

Chapter Three

Taxes

Federal income tax special considerations apply to service members. Make sure you have all of the information before filing your tax return.

IRS Publication 3, "Armed Forces' Tax Guide," addresses a wide range of issues that might affect members of the military. This should be your guide to what deductions or credits might be available to you. Common deductions are moving expenses (a portion of them), uniform upkeep, travel required to fulfill Reserve duty, and others. For service members serving in combat zones, some pay, referred to as "combat pay," may be excluded from their income.

Education credits, lifelong learning, and earned income tax credits (EITC) may also apply to you.

TIPS AND INFORMATION

The IRS website provides a collection of tax tips. These cover a range of topics, and each month a new set of tips is posted.

An app from the IRS?—yes! Released in 2011 and updated annually, the app provides information and YouTube access, and it also allows you to check the status of your refund. The app is available from the Apple App Store, or shop Google Play for the Android version.

The IRS YouTube channel features short videos with tax tips, useful information, and resources. The videos are also available in Spanish and American Sign Language.

HELP PREPARING YOUR TAXES

The military is one of the strongest partners in the Volunteer Income Tax Assistance (VITA) Program. The Armed Forces Tax Council (AFTC) consists of the tax program coordinators for the Army, Air Force, Navy, Marine Corps, and Coast Guard. The AFTC oversees the operation of the military tax programs worldwide and serves as the main conduit for outreach to military personnel and their families.

Airmen, soldiers, sailors, marines, and guardsmen and their families worldwide get tax preparation help at offices within their installations. These military VITA sites provide free tax advice, tax preparation, and assistance to military members and their families. They are trained and equipped to address military-specific tax issues, such as combat zone tax benefits and the effect of the new EITC guidelines.

Many military bases partner with VITA and provide tax help. You can contact the base office that handles family and support services; the IRS also has a VITA service locator available at http://irs.treasury.gov/freetaxprep/.

Federal taxes must be paid on all income, including wages, interest earned on bank accounts, and so on. However, some tax benefits may arise as a result of a service member serving in a combat zone, while other benefits—such as exclusions, deductions, and credits—may arise as a result of certain expenses incurred by the service member. The combat zone exclusion, for instance, is tax-free and does not have to be reported on tax returns.

The deadline for filing tax returns and paying any tax due is automatically extended for those serving and those hospitalized as a result of injuries incurred while serving in the Armed Forces in a combat zone, in a qualified hazardous duty area, or on deployment outside of the United States while participating in a contingency operation.

Combat zones are designated by an executive order from the president as areas in which the U.S. Armed Forces are engaging or have engaged in combat. There are currently three such combat zones (including the airspace above each):

Arabian Peninsula Areas, beginning January 17, 1991—the Persian Gulf, Red Sea, Gulf of Oman, the part of the Arabian Sea north of 10° North latitude and west of 68° East longitude, the Gulf of Aden, and the countries of Bahrain, Iraq, Kuwait, Oman, Qatar, Saudi Arabia, and the United Arab Emirates

Kosovo area, beginning March 24, 1999—Federal Republic of Yugoslavia (Serbia and Montenegro), Albania, the Adriatic Sea, and the Ionian Sea north of the 39th Parallel

Afghanistan, beginning September 19, 2001

In general, the deadlines for performing certain actions applicable to taxes are extended for the period of the service member's service in the combat zone, plus 180 days after the last day in the combat zone. This extension applies to the filing and paying of income taxes that would have been due April 15.

Members of the U.S. Armed Forces who perform military service in an area outside a combat zone qualify for the suspension of time provisions if their service is in direct support of military operations in the combat zone, and they receive special pay for duty subject to hostile fire or imminent danger as certified by the Department of Defense.

The deadline extension provisions apply not only to members serving in the U.S. Armed Forces (or individuals serving in support thereof) in the combat zone, but to their spouses as well, with two exceptions. First, if you are hospitalized in the United States as a result of injuries received while serving in a combat zone, the deadline extension provisions would not apply to your spouse. Second, the deadline extension provisions for a spouse do not apply for any tax year beginning more than two years after the date of the termination of the combat zone designation.

Filing individual income tax returns for your dependent children is not required while your husband is in a combat zone. Instead, these returns will be timely if filed on or before the deadline for filing your joint income tax return under the applicable deadline extensions. When filing your children's individual income tax returns, put "COMBAT ZONE" in red at the top of those returns.

TAX EXCLUSIONS: THE COMBAT ZONE EXCLUSION

If you serve in a combat zone as an enlisted person or as a warrant officer (including commissioned warrant officers) for any part of a month, all your military pay received for military service that month is excluded from gross income. For commissioned officers, the monthly exclusion is capped at the highest enlisted pay, plus any hostile fire or imminent danger pay received.

Military pay received by enlisted personnel who are hospitalized as a result of injuries sustained while serving in a combat zone is excluded from gross income for the period of hospitalization, subject to the two-year limitation provided below. Commissioned officers have a similar exclusion, limited to the maximum enlisted pay amount per month. These exclusions from gross income for hospitalized enlisted personnel and commissioned officers end two years after the date of termination of the combat zone.

Annual leave payments to enlisted members of the U.S. Armed Forces upon discharge from service are excluded from gross income to the extent that the annual leave was accrued during any month in any part of which the member served in a combat zone. If your wife is a commissioned officer, a portion of the annual leave payment she receives for leave accrued during any month in any part of which she served in a combat zone may be excluded. The annual leave payment is not excludable to the extent that it exceeds the maximum enlisted pay amount for the month of service to which it relates less the amount of military pay already excluded for that month.

The reenlistment bonus is excluded from gross income although received in a month that you were outside the combat zone, because you completed the necessary action for entitlement to the reenlistment bonus in a month during which you served in the combat zone.

A recent law change makes it possible for members of the military to count tax-free combat pay when figuring how much they can contribute to a Roth or traditional IRA. Before this change, members of the military whose earnings came from tax-free combat pay were often barred from putting money into an IRA, because taxpayers usually must have taxable earned income. Taxpayers choosing to put money into a Roth IRA don't need to report these contributions on their individual tax return. Roth contributions are not deductible, but distributions, usually after retirement, are normally tax-free. Income limits and other special rules apply.

CHANGING TAX WITHHOLDINGS

You may want to change your income tax withholdings. You can do this using *myPay* or by contacting your HR/pay office and letting them know how much you want withheld.

THRIFT SAVINGS PLAN

The Thrift Savings Plan (TSP) is a retirement savings plan with special tax advantages for federal government employees and service members. Participation is optional, and service members must join TSP while they are still serving in the military. TSP is similar to traditional 401(k) plans often sponsored by private employers; contributions to TSP accounts are not taxed at the time they are made, but distributions from the accounts generally are subject to income tax at the time the distributions are withdrawn. Veterans who did not sign up for TSP while in service cannot join the plan after leaving the military. Contributions to TSPs are subject to certain limitations. Detailed information about TSP is available in chapter 2 and at http://www.tsp.gov.

There are three basic ways for service members to access funds in their TSP accounts: in-service withdrawals, TSP loans, and post-separation withdrawals.

Regarding in-service withdrawals, most service members can borrow against the contributions and earnings made to his or her TSP account. These loans generally have no tax consequences. However, the loan must be paid back with interest, usually within five years. Payments usually take the form of payroll deductions. Therefore, service members who do not receive monthly pay (i.e., reservists with irregular training intervals) may not be eligible for TSP loans.

Service members may also withdraw money from their TSP account under what is known as a financial hardship withdrawal. Financial hardship withdrawals generally are subject to a 10 percent penalty tax, in addition to the income tax on the withdrawal. However, this 10 percent penalty generally does not apply if the withdrawal is made because of a permanent and total disability or if the money is used to pay for deductible medical expenses that exceed 7.5 percent of the service member's adjusted gross income.

The 10 percent penalty does not apply to any portion of a distribution that represents tax exempt contributions from pay earned in a combat zone. Also, combat zone exclusion pay contributed to a TSP account is not taxable when withdrawn, unlike regular pay. However, the interest earned on amounts contributed to a TSP account that were exempt from tax because of the combat zone exclusion is taxable. If a service member receives a distribution from an account that has both exclusion and non-exclusion contributions, the distribution will be paid in the same proportions as the service member's exclusion and non-exclusion contributions.

TSP participants may withdraw money from their accounts if either they are at least 59½ years old or they have a verifiable financial hardship. For instance, a disabled service member may face financial hardship in connection with his or her medical condition. In such a case, a financial hardship withdrawal may be permitted.

USMC Key Volunteer Network

USMC Key Volunteers Key Volunteer Network (KVN) is an official Marine Corps family readiness program. The network consists of Marine spouses called Key Volunteers (KVs), and they serve in both active-duty and reserve units. KVs receive formal training either from classes on base or online and are appointed by the unit commander.

The KVN structure includes a Key Volunteer Advisor (KVA) who is usually the commanding officer's spouse (or spouse of another senior officer), a Key Volunteer Coordinator (KVC) who is the executive officer's spouse (or spouse of another senior officer), and a number of additional KVs who are spouses of other Marines within the unit.

The commanding officers (CO) of individual active-duty units rely on the KVN to provide additional support and resource referrals to the Marine families of that unit. Reserve units also utilize the KVN. However, if a unit is widely geographically dispersed, the CO may appoint a parent to serve as a KV or KVC. Marine Forces Reserve parents who are local often have insight into resources and assistance that are available and helpful to unit families.

The goal of the KVN is to help families achieve and maintain family readiness. This means that they communicate official command information as directed, serve as a communication link between the command and families, and provide information to Marine families through resource referrals as needed. During deployments, the KVN is especially important because it is further utilized as a communications tool to keep families of Marines better informed about mission(s) and tasks of individual units. The Marine Corps believes that if Marines feel their families are supported and taken care of, they are better able to perform efficiently, effectively, and safely.

NEWLYWEDS

Newlyweds should seek advice about whether to change their tax status with the IRS. Depending on the date that you got married, there may be advantages in still being taxed separately. Most bases have an income tax assistance program that you can turn to for advice.

Talk over finances with your spouse. If he or she is deployed, it is important to be able to manage their finances in their absence. You can get a lot of information about pay and compensation at http://www.dod.mil/militarypay.

It is also important to familiarize yourself with your spouse's Leave and Earnings Statement (LES), which gives details about pay, allowances, and vacation time accrued.

Chapter Four

Insurance

LIFE INSURANCE

What to Look for in a Life Insurance Policy

Explore the various life insurance options. By checking each, you will be able to pick what is best for you. When shopping for life insurance, the program you choose should pay:

Funeral expenses and related bills
Debts or loans owed by the insured person at the time of death
Lost earnings

Lost earnings are what the person would have made over the rest of his or her working life had he or she not died. There are a variety of ways to calculate lost earnings. For example, the sole wage earner for a family of four dies at age forty-five. He made $30,000 a year at the time of his death. Because the household has been reduced from four to three, three-fourths (75 percent) of his income needs to be replaced for twenty years (when he would have turned age sixty-five). This method shows lost earnings that need to be recovered through insurance as $450,000: 75 percent x ($30,000 x 20 years) = $450,000.

Servicemembers Group Life Insurance (SGLI)

SGLI is a program of low-cost group life insurance for service members on active duty, Ready Reservists, members of the National Guard, members of the Commissioned Corps of the National Oceanic and Atmospheric Administration and the Public Health Service, cadets and midshipmen of the four service academies, and members of the Reserve Officer Training Corps.

SGLI coverage is available in $50,000 increments up to the maximum of $400,000. SGLI premiums are currently $.065 per $1,000 of insurance, regardless of the member's age.

Family Servicemembers' Group Life Insurance (FSGLI)

FSGLI is a program extended to the spouses and dependent children of members insured under the SGLI program. FSGLI provides up to a maximum of $100,000 of insurance coverage for spouses, not to exceed the amount of SGLI the insured member has in force, and $10,000 for dependent children. Spousal coverage is issued in increments of $10,000, at a monthly cost ranging from $.55 to $5.20 per increment.

Service members should contact their personnel support center, personnel flight, payroll, and/or finance office for SGLI and FSGLI premium payment information.

Traumatic Injury Protection under Servicemembers' Group Life Insurance (TSGLI)

The TSGLI program is a rider to SGLI. The TSGLI rider provides for payment to service members who are severely injured (on or off duty) as the result of a traumatic event and suffer a loss that qualifies for payment under TSGLI.

TSGLI payments are designed to help traumatically injured service members and their families with financial burdens associated with recovering from a severe injury. TSGLI payments range from $25,000 to $100,000 based on the qualifying loss suffered. Every member who has SGLI also has TSGLI effective December 1, 2005. TSGLI coverage is automatic for those insured under basic SGLI and cannot be declined. The only way to decline TSGLI is to decline basic SGLI coverage.

The premium for TSGLI is a flat rate of $1 per month for most service members. Members who carry the maximum SGLI coverage of $400,000 will pay $29 per month for both SGLI and TSGLI.

To be eligible for payment of TSGLI, you must meet all of the following requirements:

- You must be insured by SGLI when you experience a traumatic event.
- You must incur a scheduled loss and that loss must be a direct result of a traumatic injury.
- You must have suffered the traumatic injury prior to midnight of the day that you separate from the uniformed services.
- You must suffer a scheduled loss within two years (730 days) of the traumatic injury.

- You must survive for a period of not less than seven full days from the date of the traumatic injury. (The seven-day period begins on the date and time of the traumatic injury, as measured by Zulu [Greenwich Meridian] time, and ends 168 full hours later.)

Veterans' Group Life Insurance (VGLI)

Once your SGLI coverage extension ends, you must make your own arrangements for life insurance. One option is VGLI, offered by the VA.

VGLI provides for the conversion of SGLI to a renewable term life insurance policy. This policy is renewable every five years, regardless of health, and can be retained for life.

You are eligible to apply for VGLI if you are insured under SGLI *and*:

- You are being released from active duty or the Reserves or were released within the last year and 120 days.
- You are a member of the Individual Ready Reserve (IRR) or Inactive National Guard (ING).
- You are a reservist who suffers an injury or disability during active duty or inactive duty for training for a period of less than thirty-one days and becomes uninsurable at standard premium rates.

Remember, you can apply for VGLI within the first 120 days without evidence of good health. After the initial 120 days you have an additional year to apply, but good health requirements must be met. VGLI provides for the amount of SGLI coverage a member had in force at the time of separation from active duty or reserves. VGLI is issued in multiples of $10,000 up to a maximum 0f $400,000. VGLI can be converted at any time to an individual permanent (i.e., whole life or endowment) plan with any of the participating commercial insurance companies.

Applying for VGLI

Since SGLI coverage continues at no cost for 120 days after discharge, VGLI will not take effect until the 121st day. VGLI applications are mailed to eligible members on three occasions:

Generally within sixty days after separation.
Within 120 days after separation when the SGLI free coverage period ends.
Before the end of the sixteen-month application period.

Note: VGLI application VA Form SGLV 8714, "Application for Veterans' Group Life Insurance" (http://www.insurance.va.gov/sgliSite/forms/8714.htm), should be mailed to the address shown on your DD Form 214 or equivalent separation orders. It is your responsibility to apply within the time limits even if you do not receive an application in the mail.

Applications for VGLI coverage should be mailed to the following address:

Office of Servicemembers' Group Life Insurance
P.O. Box 5000
Millville, NJ 08332-9928

For more information, call toll-free 1-800-419-1473 or visit http://www.insurance.va.gov.

Service-Disabled Veterans Insurance (S-DVI)

S-DVI is life insurance for veterans who receive a service-connected disability rating by the Department of Veterans Affairs. The basic S-DVI program, commonly referred to as "RH Insurance," insures eligible veterans for up to $10,000 of coverage. Veterans who have the basic S-DVI coverage and are totally disabled are eligible to have their premiums waived. If waiver is granted, totally disabled veterans may apply for additional coverage of up to $20,000 under the Supplemental SDVI program. Premiums for Supplemental S-DVI coverage, however, cannot be waived.

Veterans' Mortgage Life Insurance (VMLI)

VMLI is an insurance program that provides insurance coverage on the home mortgages of veterans with severe service-connected disabilities who:

receive a Specially-Adapted Housing Grant from VA for assistance in building, remodeling, or purchasing an adapted home; *and*
have title to the home; *and*
have a mortgage on the home.

Updating Designations of Beneficiary

You may want to change your designations of beneficiary for life insurance or for retirement. New designations must be in writing and witnessed. Your agency can provide you with the appropriate forms, or you may print them from the Web. There can be no erasures or cross-outs on these forms. If the forms are incomplete or have errant marks, they will not be processed and will be returned to you.

When your beneficiary forms are filled in, be sure you have signed them. After obtaining witness signatures, submit your completed forms to the address or agency noted on the form.

Note: If you choose to complete beneficiary forms, it is your responsibility to keep them up to date. A marriage, divorce, or other change in family status does not automatically change a beneficiary form previously submitted, nor does it prevent the named beneficiary from receiving the death benefits.

If you get married and you already have a family plan, your personnel office can assist you in getting your spouse (and, if appropriate, stepchildren) added to your enrollment. Changes must be made within sixty calendar days of your marriage.

You must provide supporting documentation (e.g., marriage certificate) for any "non-open season" change. Documentation should be submitted to your HR/personnel office.

EXCEPTIONAL FAMILY MEMBER PROGRAM (EFMP)

The EFMP is a mandatory enrollment program, based on carefully defined rules. EFMP works with other military and civilian agencies to provide comprehensive and coordinated medical, educational, housing, community support, and personnel services to families with special needs. EFMP enrollment works to ensure that needed services are available at the receiving command before the assignment is made.

An exceptional family member is a dependent, regardless of age, who requires medical services for a chronic condition; receives ongoing services from a specialist; has behavioral health concerns/social problems/psychological needs; receives education services provided on an Individual Education Program (IEP); or receives services provided on an Individual Family Services Plan (IFSP).

HEALTH CARE INSURANCE PLANNING IS CRITICAL

Many service members are caught by surprise when they realize the actual cost of providing health care insurance for themselves and their families. Before you leave military service, you need to arrange for health insurance to protect you and your family. This section will help you learn about your options and plan for your health care insurance needs.

Most people leaving the military get civilian jobs that provide health care insurance. The result is continuous coverage. Sometimes, however, there is a gap between the time your service-provided coverage ends and your new employer's coverage begins. During this time, you alone are responsible for paying all the medical bills that you and your family might acquire. This could be devastating. A one-day stay in the hospital could cost thousands of dollars!

Fortunately, several resources are available to ensure continuous, comprehensive, quality health care for you and your family. Your options will be explained to you during your appointment at your Transition Office. For specific health insurance questions, call the health benefits adviser at your military medical treatment facility.

Expecting a Baby?

If you and your spouse are expecting a baby, make sure your insurance covers the infant from the date of birth, as opposed to twelve or thirteen days after birth. Medical expenses within these first two critical weeks can be costly and should be covered.

Expecting parents should meet with their local health benefits adviser early in the transition process to get additional information regarding health care and health insurance for the period following the service member's separation. Active-duty service members who separate from the military prior to delivery may be eligible to deliver the child in a military treatment facility after separation. Again, check with the commander at your military treatment facility and your health benefits adviser before you separate to see if you are eligible.

"Check Up" on Your Health before You Leave

While you are in the service, you and your family have health care coverage. The range of health care services is vast, yet your out-of-pocket expense is minimal. Use this time wisely and make an appointment early.

Get a Physical

If military treatment facilities, personnel resources, and local policy permit, you and your family members should arrange for your separation physicals as early as possible. Any problems can be treated while your medical expenses are still fully covered by the service. Take care of as much as you can prior to separation.

Get Your Records

Even if you are in good health, get a copy (certified, if possible) of your medical records from your medical treatment facility. These records will provide useful background information to the health care professionals who will assist you in your upcoming civilian life. Your military health records will be transferred (with your consent) to the VA regional office nearest your separation address.

Posttraumatic Stress Disorder (PTSD)

PTSD can occur following a life-threatening event like military combat, natural disasters, terrorist incidents, serious accidents, or violent personal assaults like rape. Most survivors of trauma return to normal given a little time. However, some people have stress reactions that don't go away on their own, or may even get worse over time. These individuals may develop PTSD.

People who suffer from PTSD often suffer from nightmares, flashbacks, difficulty sleeping, and feeling emotionally numb. These symptoms can significantly impair your daily life.

In addition, PTSD is marked by clear physical and psychological symptoms. It often has symptoms like depression, substance abuse, problems of memory and cognition, and other physical and mental health problems. The disorder is also associated with difficulties in social or family life, including occupational instability, marital problems, family discord, and difficulties in parenting. If you think you may be suffering from PTSD, the following resources and information will help you find help in dealing with PTSD and related conditions.

Online PTSD Resources

DoD Mental Health Self-Assessment (MHSA) Program This is a mental health and alcohol screening and referral program provided for military families and service members affected by deployment and mobilization. This voluntary and anonymous program is offered online, by phone, and through special events held at installations and reserve units. Anonymous self-assessments are available for depression, bipolar disorder, alcohol use, PTSD, and generalized anxiety disorder.

Individualized results and referrals to military health resources, including TRICARE, Vet Centers and Military OneSource, are provided at the end of every assessment. See http://www.pdhealth.mil/militarypathways.asp.

National Center for Posttraumatic Stress Disorder (PTSD) This special center within the Department of Veterans Affairs was created to advance the clinical care and social welfare of America's veterans through research, education, and training in the science, diagnosis, and treatment of PTSD and stress-related disorders. See http://www.ptsd.va.gov.

Ameriforce Deployment Guide This includes fact sheets and information for both service members and their families on post-deployment, including home, finances, career, and more. See http://www.ameriforce.net/deployment/.

Courage to Care This site was created by Uniformed Services University for the Health Sciences, which belongs to the Center for Traumatic Studies and includes a wealth of additional information. "Courage to Care" is an electronic health campaign for military and civilian professionals serving the military community, and for military men, women, and families. See http://www.usuhs.mil/psy/courage.html.

Military OneSource This free, twenty-four-hour service, provided by the Department of Defense, is available to all active-duty, Guard, and Reserve members and their families. Consultants provide information and make referrals on a wide range of issues. You can reach the program by telephone at 1-800-342-9647 or through the website at http://www.militaryonesource.mil.

Transitional Assistance Management Program (TAMP)

TAMP offers transitional TRICARE coverage to certain separating active-duty members and their eligible family members. Care is available for a limited time. TRICARE eligibility under the TAMP has been permanently extended to 180 days.

There are four categories of eligibility for TAMP:

- Members involuntarily separated from active duty and their eligible family members
- National Guard and Reserve members, collectively known as the Reserve Component (RC), separated from active duty after being called up or ordered in support of a contingency operation for an active-duty period of more than thirty days and their family members
- Members separated from active duty after being involuntarily retained in support of a contingency operation and their family members
- Members separated from active duty following a voluntary agreement to stay on active duty for less than one year in support of a contingency mission and their family members

Active-duty sponsors and family members enrolled in TRICARE Prime who desire to continue their enrollment upon the sponsor's separation from active-duty status are required to reenroll. To reenroll, the sponsor or family member must complete and submit a TRICARE Prime enrollment application. Contact your servicing personnel center prior to separating to see if you are TAMP eligible. Under TAMP, former active-duty sponsors, former activated reservists, and family members of both are not eligible to enroll or reenroll in TRICARE Prime Remote or in TRICARE Prime Remote for Active Duty Family Members because both programs require the sponsor to be on active duty. Under the TAMP, the sponsor is no longer on active duty and is treated as an active-duty family member for benefits and cost-sharing purposes.

Note: Transitional health care does not apply to retirees.

Once your initial transitional health care ends After this 180-day period, you and your family are no longer eligible to use military treatment facilities or TRICARE. However, you may purchase health care coverage, known as the Continued Health Care Benefit Program (CHCBP) for up to eighteen months of coverage. You have sixty days after your initial transitional health care ends to enroll in CHCBP. Your coverage will start the day after your separation.

You and your family members will be issued over-stamped identification cards that will allow you to use military treatment facilities after your separation. The cards will be marked with the dates you are eligible for transitional health care

You can learn more about TRICARE at http://www.tricare.mil/.

CHCBP: Your Option to Purchase Temporary Medical Coverage

Following the loss of eligibility to military medical benefits, you or a family member may apply for temporary, transitional medical coverage under CHCBP, a premium-based health care program providing medical coverage to a select group of former military beneficiaries. CHCBP is similar to but not part of TRICARE. The CHCBP program extends health care coverage to the following individuals when they lose military benefits:

The service member (who can also enroll his or her family members)
Certain former spouses who have not remarried
Certain children who lose military coverage

DoD contracted with Humana Military Healthcare Services to administer CHCBP. You may contact Humana Military Healthcare Services in writing or by phone for information regarding CHCBP. This includes your eligibility for enrolling in the program, to request a copy of the CHCBP enrollment application, to obtain information regarding the health care benefits that are available to CHCBP enrollees, and to obtain information regarding the premiums and out-of-pocket costs once you are enrolled.

Humana Military Healthcare Services, Inc.
Attn: CHCBP
P.O. Box 740072
Louisville, KY 40201
1-800-444-5445

The CHCBP enrollment application can also be found on the Web at http://www.humanamilitary.com/chcbp/pdf/dd2837.pdf.

CHCBP Basics

Continuous coverage. CHCBP is a health care program intended to provide you with continuous health care coverage on a temporary basis following your loss of military benefits. It acts as a "bridge" between your military health benefits and your new job's medical benefits, so you and your family will receive continuous medical coverage.

Preexisting condition coverage. If you purchase this conversion health care plan, CHCBP may entitle you to coverage for preexisting conditions often not covered by a new employer's benefit plan.

Benefits. The CHCBP benefits are comparable to the TRICARE standard benefits.

Enrollment and coverage. Eligible beneficiaries must enroll in CHCBP within sixty days following the loss of entitlement to the military health system. To enroll, you will be required to submit:

- A completed DD Form 2837, "Continued Health Care Benefit Program (CHCBP) Application."
- Documentation as requested on the enrollment form (e.g., DD Form 214, "Certificate of Release or Discharge from Active Duty"; final divorce decree; DD Form 1173, "Uniformed Services Identification and Privilege Card"). Additional information and documentation may be required to confirm an applicant's eligibility for CHCBP.
- A premium payment for the first ninety days of health coverage.

The premium rates are approximately $930 per quarter for individuals and $2,000 per quarter for families.

Humana Military Healthcare Services will bill you for subsequent quarterly premiums through your period of eligibility once you are enrolled.

The program uses existing TRICARE providers and follows most of the rules and procedures of the TRICARE standard program.

Depending on your beneficiary category, CHCBP coverage is limited to either eighteen or thirty-six months as follows:

eighteen months for separating service members and their families
thirty-six months for others who are eligible (in some cases, former spouses who have not remarried may continue their coverage beyond thirty-six months if they meet certain criteria)

You may not select the effective date of coverage under CHCBP. For all enrollees, CHCBP coverage must be effective on the day after you lose military benefits.

For more information about CHCBP, visit their website at http://www.humanamilitary.com/south/bene/TRICAREPrograms/chcbp.asp or call their toll-free line at 1-800-444-5445.

Chapter Five

Home Owning

Everyone in the military gets free (or almost free) housing. How the military chooses to provide this to you depends mostly upon your marital (dependency) status or your rank. If you are married and living with your spouse and/or minor dependents, you will either live in on-base housing or be given a monetary allowance called Basic Allowance for Housing (BAH) to live off-base. The amount of BAH is dependent upon your rank, your location, and whether or not you have dependents.

If you are in the Guard or Reserves and entitled to a housing allowance, you will receive a special, reduced BAH, called BAH Type II, any time you are on active duty for less than thirty days. If you are on orders to serve on active duty for thirty days or more, you'll receive the full housing allowance rate (the same as active duty).

If you have dependents, you will receive the housing allowance even when staying in the barracks at basic training and/or technical school/AIT/A-school. This is because the military makes it mandatory for you to provide adequate housing for your dependents. This will be included as part of your regular paycheck. (*Note*: In the military, your monthly pay entitlements are paid twice per month—half on the first of the month, and half on the last duty day of the month.) For basic training and/or technical school/AIT/A-school, you will receive the BAH amount for the location where your dependent(s) are residing.

However, if you are not married and/or divorced and are paying child support, you do not receive full-rate BAH while living in the barracks. In this case, special rules apply, and the member receives BAH-DIFF. Unlike basic pay, BAH is an "allowance," not a "pay," and is therefore not taxable.

If you are single, you can expect to spend the few years of your military service residing on-base in the dormitory, or "barracks." Policies concerning single military members living off-base at government expense vary from service to service, and even from base to base, depending on the occupancy rate of the barracks/dormitories on the particular base.

Army policy allows single members in the pay grade of E-6 and above to live off-base at government expense. However, at some bases, E-5s are allowed to move off-base at government expense, depending on the barracks occupancy rates of that base.

The Air Force policy generally allows single E-4s with more than three years of service, and above, to reside off-base at government expense.

The Navy policy allows single sailors in the pay grades of E-5 and above, and E-4s with more than four years of service to reside off-base and receive a housing allowance.

The Marines allow single E-6s and above to reside off-base at government expense. On some bases, depending on the barracks occupancy rate, single E-5s and even some E-4s are authorized to reside off-base.

DORMITORIES

If your recruiter promised you condos, you're out of luck. (Remember the movie *Private Benjamin*?) However, all of the services have implemented plans to improve single housing (dormitories/barracks) for enlisted personnel.

The Air Force was the first service to get started on the program, and they are arguably ahead of the other services. All airmen, outside of basic training and technical school, are now entitled to a private room. The Air Force started with remodeling barracks into a concept called one-plus-one, which provided a private room, a small kitchen, and a bathroom/shower shared with one other person. The Air Force has now upgraded their program using a concept called "Dorms-4-Airmen." All new Air Force dormitories (except basic training and technical school) are now designed using this concept. Dormitories under this program are four-bedroom apartments. Airmen have a private room and private bath and share a kitchen, washer and dryer, and living room with three other airmen.

The Army's standard is a two-bedroom apartment designed for two soldiers. Each soldier gets a private bedroom and shares a kitchen, bathroom, and living room.

The Navy had a serious problem when this initiative started. Thousands of their junior sailors were living on ships, even when their assigned ships were in port. Constructing enough barracks on Navy bases to provide single rooms for all of these sailors would cost a fortune. The Navy solved this problem by getting permission from Congress to use private industry to construct and operate privatized housing for lower-ranking single sailors.

Like that of the Army, this design is a two-bedroom apartment. Each sailor will have a private bedroom, a private bathroom, and share a kitchen, dining area, and living room with another sailor. However, under the Navy's Home-port Ashore initiative, sailors assigned to ships that are in port must share a bedroom until additional funding becomes available to build new complexes. Like privatized family housing, the sailor will pay the complex management monthly rent (which is equal to their housing allowance). The "rent" covers all utilities and rental insurance. The plan calls for the apartment complexes to include fitness facilities, media centers, Wi-Fi lounges, and technology centers. The first two contracts were for the Navy's two largest fleet areas (San Diego, California, and Norfolk, Virginia) and both are now in use. Similar projects are currently under way at Hampton Roads, Virginia, and Mayport, Florida. The aim is to provide accommodation for all junior unaccompanied sailors by 2016.

The Marines have taken a different route. The Marine Corps believes that lower-ranking enlisted Marines living together is essential to discipline, unit cohesion, and esprit de corps. Under the Marine Corps program, junior Marines (E-1 to E-3) share a room and a bathroom. Marines in dormitory rooms are normally subject to two types of inspections: First, there is the normal or periodic inspection which may or may not be announced in advance. This is when the commander or first sergeant (or other designated person) inspects your room to make sure you are abiding by the standards (bed made, trash empty, room clean, etc.). The second type of inspection is called a Health and Welfare Inspection (HWI). This type of inspection is always unannounced, often occurs about 2:00 a.m., and is comprised of an actual search of the dormitory rooms for contraband (drugs, guns, knives, etc.). At times, these HWIs are accompanied by a random urinalysis test, looking for evidence of drug abuse.

Some services/bases allow you to use your own furniture. Others are very strict about using only the provided government furniture. Even if you are required to use government furniture, you can have your own stereo, television, or computer system.

All in all, most single enlisted people look forward to the day when they can move out of the dormitory.

MOVING OUT

At most locations, single members can elect to move out of the dormitory and get a place off-base at their own expense. That means the government will not give them BAH, nor will the government give them a food allowance. Unless you get a roommate (or two), it can be hard to make ends meet, living off-base, with just your base pay.

By law, the services cannot allow single members to move off-base at government expense, unless the base-wide dormitory occupancy rate exceeds 95 percent. This means that over 95 percent of all dormitory rooms on the base must have people living in them before anyone can be allowed to move out of the dormitories and receive a housing allowance.

Unfortunately, dormitory/barracks spaces are usually allocated to specific units (squadrons, divisions, companies, etc.), and commanders are notoriously against allowing members of their units to live in other units' dormitories/barracks. Therefore, it's entirely possible for your particular unit to be overcrowded in the dormitory (thereby mandating that you have a roommate), while another unit has plenty of space. Unfortunately, unless the *base-wide* occupancy rate exceeds 95 percent, your commander can't authorize you to move off-base at government expense.

When the base-wide occupancy rate does exceed 95 percent, the way it is usually done is that the base offers the chance to move off-base to dormitory residents, based on rank. That is, the person (base-wide) with the most rank is offered the chance to move out first, followed by the person (base-wide) who has the next most rank, and so on, until the base-wide occupancy rate falls below 95 percent.

That means it's entirely possible that your particular dorm may be overcrowded, but the person given the chance to move off-base may be in another dorm, which is not nearly so crowded, and there you are—stuck with a roommate because your commander won't let you move to another unit's uncrowded dormitory. The solution to this problem is to periodically reallocate dormitory spaces, but this is a major hassle, and most bases, in my experience, are reluctant to tackle the project any more often than every five years or so. This mismanaged system is the source of more frustration among single military members than any other factor of military life that we're aware of.

ON-BASE HOUSING

Most places have limited on-base housing, so there is usually a waiting list (sometimes more than one year!). To qualify for on-base housing, you must be residing with a dependent (in most cases, that means spouse or minor children). The number of bedrooms you'll be authorized depends on the number and age of the dependents residing with you. Some bases have very, very nice housing; on other bases the housing barely qualifies for slum status. Utilities (trash, water, gas, electric) are normally free. Cable TV and phones are not. Furniture is normally not provided (although many bases have "loan closets," which will temporarily lend you furniture). Appliances, such as stoves and refrigerators, are usually provided. Many on-base houses even have dishwashers.

Clothes washers and dryers are usually not provided, but most units—at least in the states—have hookups. Additionally, many bases have laundromats located close to the housing area. Overseas, many housing units are condo-style, and there is a laundry room with washers and dryers located in each stairwell.

GOVERNMENT FAMILY HOUSING

The insides of occupied housing units are not normally inspected as dormitories are (although they may be inspected no-notice if the commander receives any reports of safety or sanitary problems). The outside of housing is an entirely different matter. All of the services are pretty strict about dictating exactly how the outside of the house (yard) will be maintained. Most of them employ personnel who drive by each and every housing unit once per week, and write tickets for any discrepancies noted. Receive too many tickets in too short a period of time, and you will be requested to move off-base.

In the states, most on-base family housing units are duplexes or sometimes fourplexes. For officers and more senior enlisted members, on-base family housing in the states is usually either duplexes or single dwellings. Sometimes there are fenced-in back yards, and at other bases there are not. Usually, if the housing unit has a back yard but no fence, you can get permission to install a fence at your own expense (you have to agree to take the fence down when you move out if the next occupant decides he or she doesn't want a fence).

The same is true of almost any improvement you wish to make to on-base family housing. Usually you can get permission to do self-help improvements, but you must agree to return the house to its original state if the next person to move in doesn't want to accept your improvement.

Overseas, on-base family housing units are generally in the form of high-rise apartment buildings—kind of like condominiums.

Moving out of base housing is a lot harder than moving in. This is the one time when the inside of the house *will* be inspected, and it will be expected to be in immaculate condition (many people hire professional cleaners prior to checkout). However, many bases now have programs where the base itself hires professional cleaners when an occupant moves out, making the process much easier.

More and more military bases are moving to privatized family housing. This housing is maintained, managed (and sometimes built) by private industry. The "rent" for these privatized units is paid to the housing management agency by military pay allotment, and it is equal to the member's housing allowance.

OFF-BASE HOUSING

Instead of living in the dormitories or residing in on-base housing, you may be authorized to live off-base. In this case, the military will pay you BAH. The amount of this nontaxable allowance is dependent upon your rank, marital (dependency) status, and the area you (or your dependents) live in. Once per year, the military hires an independent agency to survey the average housing costs in all of the areas where significant numbers of military personnel live. The Per Diem, Travel and Transportation Allowance Committee uses this data to compute the amount of BAH you will receive each month. (This amount is currently designed to cover 100 percent of average housing costs.The latest survey of housing costs for all fifty states can be found at http://www.defensetravel.dod.mil/site/pdcFiles.cfm?dir=/Allowances/.)

One of the nice features about the BAH law is that the amount of BAH you receive may never go down while you are living in an area, even if the average cost of housing in that area goes down. Of course, once you move to a different base, your BAH will be recalculated for the current rate in the new location.

An interesting aspect of BAH is the type of housing that the entitlement is based upon. BAH is based on acceptable housing for an individual (or an individual with dependents). For example, a married E-5 is reimbursed based on what DoD considers minimum acceptable housing, a two-bedroom townhouse or duplex. For an O-5, it is a four-bedroom detached home. While whether or not one has dependents is a factor, the number of dependents is not. See http://www.defensetravel.dod.mil/site/bah.cfm for more information.

If you move into off-base housing overseas, your monthly entitlement is called Overseas Housing Allowance (OHA), and it is recalculated every two weeks. This is because currency rates can fluctuate dramatically overseas, causing housing expenses to go up and down. In addition to OHA, those overseas are entitled to some additional allowances, such as an initial move-in expense allowance and reimbursement for costs to improve the security of the off-base residence.

If you are authorized to reside off-base, it's very important that you ensure your lease contains a "military clause." A military clause allows you to break your lease in case you are forced to move on official orders.

SPECIAL CONSIDERATIONS

If you are married to a nonmilitary member and/or you have children, your spouse and children are considered to be "dependents" by the military.

The military requires you to provide adequate support (which includes housing) to your dependents. Because of this, if you are married, you receive a housing allowance at the "with dependent" rate, even if you are living in the single dormitories/barracks.

Because living in the barracks/dormitories is mandatory during basic training and job-school, and because your dependents are not allowed to travel to basic training and/or job-school (unless the job-school is over twenty weeks long at a single location) at government expense, during these periods you live in the barracks/dormitories, and you receive BAH for the area where your dependents reside.

When you move to your first permanent duty station, the rules change. Your dependents are allowed to move there at government expense. If they don't move there, that is considered your choice. In such cases, you receive BAH (at the "with dependent" rate) for the area of your duty station, regardless of where your dependent is actually living.

As long as you are still married, to give up BAH, you would have to reside in on-base family housing. However, unless your dependents move to your duty location, you are not authorized to reside in on-base family housing, because the rules say that to qualify, your dependents must be living with you.

If there is extra space available in the barracks/dormitories, you are allowed to live there and still receive your BAH. However, now that the military is trying to give all single people living in the dormitories their own room, most bases do not have any extra space available in their dormitories. Therefore, as a married person who has voluntarily elected not to be accompanied by your dependents, you will likely be required to live off-base. You will receive BAH for the area you are assigned to. If you are allowed to live in the dormitory/barracks space available, you must be prepared to move out, with little or no notice, in case the space is needed (although most commanders/first sergeants will try to give at least two weeks' notice, if possible).

The rules change for overseas assignments. If you are assigned overseas and elect not to be accompanied by your dependents, you can live in the barracks/dormitories on base and still receive BAH in order to provide adequate housing support in the states for your dependents.

BUYING YOUR FIRST HOME

Buying your own home is always a sound investment but it can be difficult if you frequently have to relocate. It is also a major investment and one that many young couples have to think seriously about. However, with the present housing market there has never been a better time to buy, and as a military couple you are eligible for special mortgage rates.

There are a lot of things that have to be considered. Buying a home is not only a big investment, it is an ongoing one, with upkeep and maintenance, property taxes, and other fees and services that have to be paid. There is no doubt, however, that if you can afford to buy a house now and perhaps rent it out if you are posted elsewhere, it will steadily appreciate in value and be a valuable nest egg in years to come.

Some of the advantages of buying your own home are that you save on your income tax because your mortgage interest payments are deductible every year, and you own the place so you are not paying rent every month with nothing back in return. If you own your home, you can customize it to your liking—something most landlords don't encourage, and anyway, why would you spend money on upgrading a property you do not own? Another major advantage is that your house will appreciate in value, and as it does so, your equity in the property will increase. In a few years if you need extra cash—to start a family, for instance—you can take out a loan against that equity.

There are downsides as well. Buying a home is an expensive initial investment, although buyers can exert a lot of leverage at the moment, such as getting the seller to pay all or some of your closing costs. If the seller is anxious to sell, this might be acceptable to him or her in order to close the deal. There are other costs such as legal fees, surveys, deposits, and so on. There are then the ongoing payouts for mortgage, taxes, and homeowners' insurance.

If you buy you will become responsible for upkeep and maintenance—if something goes wrong, you have to fix it. There may be housing association fees that have to be paid.

To buy or not to buy has to be a decision taken together after much thought. Do you both want to be tied down with home ownership and everything that goes with it? Can you afford to buy a property—do you earn enough to pay the mortgage every month and not get into arrears?

If you expect to get a new posting in a couple of years, it is unlikely that the house will have appreciated sufficiently for you to recoup your buying costs, and it is probably better to rent. If you think you are not going to move for four or five years, then buying becomes a serious option.

If you buy a house near a base and are relocated, it may be easier to rent your home to another military couple rather than sell it. If you plan on selling each time you move on, you can always move into rented accommodations on a short lease rather than make a snap decision to buy after only a short time of looking at what is available. Rushed decisions are seldom the right ones.

Things to Consider

- Consider your ability to pay (or afford) a loan on a home. You will need to determine the maximum amount you are eligible to borrow, the length of time you want to pay on the loan, the amount of interest the lender will charge, and the insurance and other fees related to a home loan. You can use an online calculator to determine the maximum loan amount for which you can qualify.
- Know your income, expenses, and all debts. You'll need to reveal them to ensure your financial capability to successfully handle the loan, reducing the risk to you and the lender.
- As in a car loan, you should get a copy of your credit report and fix any errors.
- Consider the terms of the loan, as well as any penalties for being late on payment or prepayment fees.
- Consider the services of a realtor and mortgage broker to find your ideal home and how to finance it. They can assist you in other factors, such as home inspections, homeowners associations and their fees, taxes, insurances, escrow accounts, and other matters relevant to a home purchase.

Frequently Asked Questions

Q. Why should I buy, instead of rent?

A home is an investment. When you rent, you write your monthly check and that money is gone forever. But when you own your home, you can deduct the cost of your mortgage loan interest from your federal income taxes and usually from your state taxes. This will save you a lot each year, because the interest you pay will make up most of your monthly payment for most of the years of your mortgage. You can also deduct the property taxes you pay as a homeowner. In addition, the value of your home may go up over the years. Finally, you'll enjoy having something that's all yours—a home where your own personal style will tell the world who you are.

Q. What are "HUD homes," and are they a good deal?

HUD homes can be a very good deal. When someone with a HUD-insured mortgage can't meet the payments, the lender forecloses on the home; HUD pays the lender what is owed; and HUD takes ownership of the home. Then HUD sells it at market value as quickly as possible.

Q. Can I become a home buyer even if I have bad credit and don't have much for a down payment?

You may be a good candidate for one of the federal mortgage programs. Start by contacting one of the HUD-funded housing counseling agencies that can help you sort through your options. Also, contact your local government to see if there are any local home-buying programs that might work for you. Look in the blue pages of your phone directory for your local office of housing and community development or, if you can't find it, contact your mayor's office or your county executive's office.

Q. Should I use a real estate broker? How do I find one?

Using a real estate broker is a very good idea. All the details involved in home buying, particularly the financial ones, can be mind-boggling. A good real estate professional can guide you through the entire process and make the experience much easier. A real estate broker will be well acquainted with all the important things you'll want to know about a neighborhood you may be considering—the quality of schools, the number of children in the area, the safety of the neighborhood, traffic volume, and more. He or she will help you figure the price range you can afford and search the classified ads and multiple listing services for homes you'll want to see. With immediate access to homes as soon as they're put on the market, the broker can save you hours of wasted driving-around time. When it's time to make an offer on a home, the broker can point out ways to structure your deal to save you money. He or she will explain the advantages and disadvantages of different types of mortgages, guide you through the paperwork, and be there to hold your hand and answer last-minute questions when you sign the final papers at closing. And you don't have to pay the broker anything! The payment comes from the home seller—not from the buyer.

By the way, if you want to buy a HUD home, you will be required to use a real estate broker to submit your bid. To find a broker who sells HUD homes, check your local yellow pages or the classified section of your local newspaper.

Q. How much money will I have to come up with to buy a home?

Well, that depends on a number of factors, including the cost of the house and the type of mortgage you get. In general, you need to come up with enough money to cover three costs: *earnest money*, the deposit you make on the home when you submit your offer, to prove to the seller that you are serious about wanting to buy the house; the *down payment*, a percentage of the cost of the home that you must pay when you go to settlement; and *closing costs*, the costs associated with processing the paperwork to buy a house.

When you make an offer on a home, your real estate broker will put your earnest money into an escrow account. If the offer is accepted, your earnest money will be applied to the down payment or closing costs. If your offer is not accepted, your money will be returned to you. The amount of your earnest money varies. If you buy a HUD home, for example, your deposit generally will range from $500 to $2,000.

The more money you can put into your down payment, the lower your mortgage payments will be. Some types of loans require 10–20 percent of the purchase price. That's why many first-time homebuyers turn to HUD's FHA for help. FHA loans require only 3 percent down—and sometimes less.

Closing costs—which you will pay at settlement—average 3–4 percent of the price of your home. These costs cover various fees your lender charges and other processing expenses. When you apply for your loan, your lender will give you an estimate of the closing costs, so you won't be caught by surprise. If you buy a HUD home, HUD may pay many of your closing costs.

Q. How do I know if I can get a loan?

If the amount you can afford is significantly less than the cost of homes that interest you, then you might want to wait awhile longer. But before you give up, why don't you contact a real estate broker or a HUD-funded housing counseling agency? They will help you evaluate your loan potential. A broker will know what kinds of mortgages lenders are offering and can help you choose a lender with a program that might be right for you. Another good idea is to get prequalified for a loan. That means you go to a lender and apply for a mortgage before you actually start looking for a home. Then you'll know exactly how much you can afford to spend, and it will speed the process once you do find the home of your dreams.

Q. How do I find a lender?

You can finance a home with a loan from a bank, a savings and loan, a credit union, a private mortgage company, or various state government lenders. Shopping for a loan is like shopping for any other large purchase: you can save money if you take some time to look around for the best prices. Different lenders can offer quite different interest rates and loan fees, and, as you know, a lower interest rate can make a big difference in how much home you can afford. Talk with several lenders before you decide. Most lenders need three to six weeks for the whole loan approval process. Your real estate broker will be familiar with lenders in the area and what they're offering. Or you can look in your local newspaper's real estate section—most papers list interest rates being offered by local lenders. You can find FHA-approved lenders in the yellow pages of your phone book. HUD does not make loans directly—you must use a HUD-approved lender if you're interested in an FHA loan.

Q. In addition to the mortgage payment, what other costs do I need to consider?

Well, of course you'll have your monthly utilities. If your utilities have been covered in your rent, this may be new for you. Your real estate broker will be able to help you get information from the seller on how much utilities normally cost. In addition, you might have homeowner association or condo association dues. You'll definitely have property taxes, and you also may have city or county taxes. Taxes normally are rolled into your mortgage payment. Again, your broker will be able to help you anticipate these costs.

Q. What will my mortgage cover?

Most loans have four parts: *principal*, the repayment of the amount you actually borrowed; *interest*, payment to the lender for the money you've borrowed; *homeowners insurance*, a monthly amount to insure the property against loss from fire, smoke, theft, and other hazards, required by most lenders; and *property taxes*, the annual city/county taxes assessed on your property, divided by the number of mortgage payments you make in a year. Most loans are for thirty years, although fifteen-year loans are available, too. During the life of the loan, you'll pay far more in interest than you will in principal—sometimes two or three times more! Because of the way loans are structured, in the first years you'll be paying mostly interest in your monthly payments. In the final years, you'll be paying mostly principal.

Q. What do I need to take with me when I apply for a mortgage?

Good question! If you have everything with you when you visit your lender, you'll save a good deal of time. You should have: (1) Social Security numbers for both you and your spouse, if both of you are applying for the loan; (2) copies of your checking and savings account statements for the past six months; (3) evidence of any other assets like bonds or stocks; (4) a recent paycheck stub detailing your earnings; (5) a list of all credit card accounts and the approximate monthly amounts owed on each; (6) a list of account numbers and balances due on outstanding loans, such as car loans; (7) copies of your last two years' income tax statements; and (8) the name and address of someone who can verify your employment. Depending on your lender, you may be asked for other information.

Q. I know there are lots of types of mortgages—how do I know which one is best for me?

You're right—there are many types of mortgages, and the more you know about them before you start, the better. Most people use a fixed-rate mortgage. In a fixed-rate mortgage, your interest rate stays the same for the term of the mortgage, which normally is thirty years. The advantage of a fixed-rate mortgage is that you always know exactly how much your mortgage payment will be, and you can plan for it. Another kind of mortgage is an adjustable rate mortgage (ARM). With this kind of mortgage, your interest rate and monthly payments usually start lower than a fixed-rate mortgage. But your rate and payment can change either up or down, as often as once or twice a year. The adjustment is tied to a financial index, such as the U.S. Treasury Securities index. The advantage of an ARM is that you may be able to afford a more expensive home because your initial interest rate will be lower. There are several government mortgage programs, including the Veteran's Administration programs. Most people have heard of FHA mortgages. FHA doesn't actually make loans. Instead, it insures loans so that if buyers default for some reason, the lenders will get their money. This encourages lenders to give mortgages to people who might not otherwise qualify for a loan. Talk to your real estate broker about the various kinds of loans, before you begin shopping for a mortgage.

Q. When I find the home I want, how much should I offer?

Again, your real estate broker can help you here. But there are several things you should consider: (1) Is the asking price in line with prices of similar homes in the area? (2) Is the home in good condition or will you have to spend a substantial amount of money making it the way you want it? You probably want to get a professional home inspection before you make your offer. Your real estate broker can help you arrange one. (3) How long has the home been on the market? If it's been for sale for a while, the seller may be more eager to accept a lower offer. (4) How much mortgage will be required? Make sure you really can afford whatever offer you make. (5) How much do you really want the home? The closer you are to the asking price, the more likely your offer will be accepted. In some cases, you may even want to offer more than the asking price, if you know you are competing with others for the house.

Q. What if my offer is rejected?

They often are! But don't let that stop you. Now you begin negotiating. Your broker will help you. You may have to offer more money, but you may ask the seller to cover some or all of your closing costs or to make repairs that wouldn't normally be expected. Often, negotiations on a price go back and forth several times before a deal is made. Just remember—don't get so caught up in negotiations that you lose sight of what you really want and can afford!

Q. So, what will happen at closing?

Basically, you'll sit at a table with your broker, the broker for the seller, probably the seller, and a closing agent. The closing agent will have a stack of papers for you and the seller to sign. While he or she will give you a basic explanation of each paper, you may want to take the time to read each one and/or consult with your agent to make sure you know exactly what you're signing. After all, this is a large amount of money you're committing to pay for a lot of years! Before you go to closing, your lender is required to give you a booklet explaining the closing costs, a "good faith estimate" of how much cash you'll have to supply at closing, and a list of documents you'll need at closing. If you don't get those items, be sure to call your lender *before* you go to closing. Don't hesitate to ask questions.

DEBT-TO-INCOME RATIO

Calculating your debt-to-income ratio is as simple as adding up all of your debt and subtracting it from your income. Some calculations may exclude things like mortgage payments and property taxes, but to really get a complete picture it's best to include everything.

To determine your debt-to-income ratio, simply take your total debt payment number and divide it by your total monthly income.

Example: If you came up with a $2,000 total debt payment number and monthly income of $6,000, that leaves you with a debt-to-income ratio of 33 percent.

Lenders tend to look at two key debt-to-income ratios when it comes to mortgages:

The front ratio: the debt-to-income ratio that includes all housing costs
The back ratio: the nonmortgage debt-to-income ratio

Generally speaking, lenders would like to see your front ratio at 36 percent or less and your back ratio at 28 percent or less. The FHA rates are 41 percent and 29 percent, front and back, respectively. Keep in mind that these ratios are only guidelines, and there are many other factors that go into determining how much you can borrow and at what rate.

CREDIT REPORTS

Nationwide consumer reporting companies sell the information in your credit report (see also chapter 2) to creditors, insurers, employers, and other businesses that use it to evaluate your applications for credit, insurance, employment, or renting a home. The Fair Credit Reporting Act requires each of the three nationwide consumer reporting companies (Equifax, TransUnion, and Experian) to provide you with a free copy of your credit report, at your request, once every twelve months.

A credit report includes information on where you live, how you pay your bills, and whether you've been sued or arrested, or have filed for bankruptcy. Because nationwide consumer reporting companies get their information from different sources, the information in your report from one company may not reflect all, or the same, information in your reports from the other two companies. That's not to say that the information in any of your reports is necessarily inaccurate; it just may be differently interpreted.

YOUR CREDIT SCORE AND WHAT IT MEANS

A credit score is a three-digit number that lenders use to help them decide whether you're a reliable candidate for a mortgage, credit card, or some other line of credit; it also helps determine the interest rate you are charged for this credit. Fair Isaac has developed a unique scoring system for each of the three credit bureaus, taking the following five components into account:

Payment history (35 percent)
How much you owe (30 percent)
Length of credit history (15 percent)
Type of credit (10 percent)
New credit or inquiries (10 percent)

Scores range from approximately 300 to 850. The higher your score, the better the terms of credit you are likely to receive. The riskier you appear to the lender—or the lower the score—the less likely you will get credit or, if you are approved, the more that credit will cost you. In other words, you will pay more to borrow money.

HOME EQUITY LINE OF CREDIT (HELOC)

A home equity line of credit is a form of revolving credit in which your home serves as collateral. Because a home often is a consumer's most valuable asset, many homeowners use home equity credit lines only for major items, such as education, home improvements, or medical bills, and choose not to use them for day-to-day expenses.

With a home equity line, you will be approved for a specific amount of credit. Many lenders set the credit limit on a home equity line by taking a percentage (say, 75 percent) of the home's appraised value and subtracting from that the balance owed on the existing mortgage.

In determining your actual credit limit, the lender will also consider your ability to repay the loan (principal and interest) by looking at your income, debts, and other financial obligations, as well as your credit history. Home equity lines of credit typically involve variable rather than fixed interest rates. The variable rate must be based on a publicly available index (such as the prime rate published in some major daily newspapers or a U.S. Treasury bill rate). In such cases, the interest rate you pay for the line of credit will change, mirroring changes in the value of the index.

Remember that your home is at risk when you take out a HELOC—use the line of credit only when you must, and be sure you have the ability to repay it.

HOME INSURANCE

Lending institutions usually require mortgage customers to purchase home insurance, and they suggest certain coverage limits that must be maintained by the mortgagee. Home insurance includes the structure, the contents (your possessions), and liability should someone get injured on your property. Before buying home insurance, understand the difference between "replacement cost" and "actual cash value." Actual cash value is an item's replacement cost, minus depreciation. Replacement cost is how much it would take to replace the item or home without depreciation. Extended replacement cost coverage pays a certain amount above the policy limit to replace a damaged home, generally 120 or 125 percent.

What You Need to Know When Deciding Coverage

Don't rely on the coverage levels mandated by your bank or mortgage company. Those levels are designed to protect the house itself, but not necessarily your possessions.

Insure your house for the cost to replace it (i.e., construction costs), not its market value, and don't factor in the value of your land. Once you know the proper level of coverage, consider special add-ons for valuables such as jewelry, your computer equipment, and other pricey possessions.

You might also need additional coverage for earthquakes, flooding, or windstorms, depending on where you live. Each homeowners insurance policy provides a combination of property and liability coverage and covers loss of use resulting from damage.

Factors That Go into Determining the Premiums for a Homeowners Policy

The age of your home, the materials used to build it, where it's located, the square footage, and its distance from a fire hydrant all play a role in determining rates. The insurer will be able to give you an estimate for rebuilding your house in the event of a total loss.

MORTGAGE PAYMENTS SENDING YOU REELING?

The possibility of losing your home because you can't make the mortgage payments can be terrifying. Perhaps you are one of the many consumers who took out a mortgage that had a fixed rate for the first two or three years and then had an adjustable rate. Or maybe you're anticipating an adjustment, and you want to know what your payments will be and whether you'll be able to make them. Or maybe you're having trouble making ends meet because of an unrelated financial crisis.

Regardless of the reason for your mortgage anxiety, the Federal Trade Commission (FTC), the nation's consumer protection agency, wants you to know how to help save your home and how to recognize and avoid foreclosure scams.

Know Your Mortgage

Do you know what kind of mortgage you have? Do you know whether your payments are going to increase? If you can't tell by reading the mortgage documents you received at settlement, contact your loan servicer and ask. (A loan servicer is responsible for collecting your monthly loan payments and crediting your account.)

Here are some examples of types of mortgages:

- *Hybrid adjustable rate mortgages (ARMs).* Mortgages that have fixed payments for a few years, and then turn into adjustable loans. Some are called 2/28 or 3/27 hybrid ARMs: the first number refers to the years the loan has a fixed rate and the second number refers to the years the loan has an adjustable rate. Others are 5/1 or 3/1 hybrid ARMs: the first number refers to the years the loan has a fixed rate, and the second number refers to how often the rate changes. In a 3/1 hybrid ARM, for example, the interest rate is fixed for three years, then adjusts every year thereafter.

- *ARMs*: Mortgages that have adjustable rates from the start, which means your payments change over time.
- *Fixed-rate mortgages*: Mortgages where the rate is fixed for the life of the loan; the only change in your payment would result from changes in your taxes and insurance if you have an escrow account with your loan servicer.

If you have a hybrid ARM or an ARM and the payments will increase and you have trouble making the increased payments, find out if you can refinance to a fixed-rate loan. Review your contract first, checking for prepayment penalties. Many ARMs carry prepayment penalties that force borrowers to come up with thousands of dollars if they decide to refinance within the first few years of the loan. If you're planning to sell soon after your adjustment, refinancing may not be worth the cost. But if you're planning to stay in your home for a while, a fixed-rate mortgage might be the way to go. Online calculators can help you determine your costs and payments.

If You Are Behind on Your Payments

If you are having trouble making your payments, contact your loan servicer to discuss your options as early as you can. Most loan servicers are willing to work with customers they believe are acting in good faith, and those who call them early on. The longer you wait to call, the fewer options you will have. After you've missed three or four payments and your loan is in default, most loan servicers won't accept a partial payment of what you owe. They will start foreclosure unless you can come up with the money to cover all your missed payments, plus any late fees.

Contacting Your Loan Servicer

Before you have any conversation with your loan servicer, prepare. Record your income and expenses, and calculate the equity in your home. To calculate the equity, estimate the market value less the balance of your first and any second mortgage or home equity loan. Then write down the answers to the following questions:

What happened to make you miss your mortgage payment(s)? Do you have any documents to back up your explanation for falling behind? How have you tried to resolve the problem?

Is your problem temporary, long-term, or permanent? What changes in your situation do you see in the short term, and in the long term? What other financial issues may be stopping you from getting back on track with your mortgage?

What would you like to see happen? Do you want to keep the home? What type of payment arrangement would be feasible for you?

Throughout the foreclosure prevention process, keep notes of all your communications with the servicer, including date and time of contact, the nature of the contact (face-to-face, by phone, e-mail, fax, or postal mail), the name of the representative, and the outcome.

Follow up any oral requests you make with a letter to the servicer. Send your letter by certified mail, return receipt requested, so you can document what the servicer received. Keep copies of your letter and any enclosures.

Meet all deadlines the servicer gives you.

Avoiding Default and Foreclosure

If you have fallen behind on your payments, consider discussing the following foreclosure prevention options with your loan servicer:

Reinstatement. You pay the loan servicer the entire past-due amount, plus any late fees or penalties, by a date you both agree to. This option may be appropriate if your problem paying your mortgage is temporary.

Repayment plan. Your servicer gives you a fixed amount of time to repay the amount you are behind by adding a portion of what is past due to your regular payment. This option may be appropriate if you've missed only a small number of payments.

Forbearance. Your mortgage payments are reduced or suspended for a period you and your servicer agree to. At the end of that time, you resume making your regular payments as well as a lump sum payment or additional partial payments for a number of months to bring the loan current. Forbearance may be an option if your income is reduced temporarily (for example, you are on disability leave from a job, and you expect to go back to your full-time position shortly). Forbearance isn't going to help you if you're in a home you can't afford.

Loan modification. You and your loan servicer agree to permanently change one or more of the terms of the mortgage contract to make your payments more manageable for you. Modifications can include lowering the interest rate, extending the term of the loan, or adding missed payments to the loan balance. A loan modification may be necessary if you are facing a long-term reduction in your income.

Before you ask for forbearance or a loan modification, be prepared to show that you are making a good-faith effort to pay your mortgage. For example, if you can show that you've reduced other expenses, your loan servicer may be more likely to negotiate with you.

Selling your home. Depending on the real estate market in your area, selling your home may provide the funds you need to pay off your current mortgage debt in full.

Bankruptcy. Personal bankruptcy generally is considered the debt management option of last resort because the results are long-lasting and far-reaching. A bankruptcy stays on your credit report for ten years and can make it difficult to obtain credit, buy another home, get life insurance, or sometimes even get a job. Still, it is a legal procedure that can offer a fresh start for people who can't satisfy their debts.

If you and your loan servicer cannot agree on a repayment plan or other remedy, you may want to investigate filing Chapter 13 bankruptcy. If you have a regular income, Chapter 13 may allow you to keep property like a mortgaged house or car that you might otherwise lose. In Chapter 13, the court approves a repayment plan that allows you to use your future income toward payment of your debts during a three-to-five-year period, rather than surrender the property. After you have made all the payments under the plan, you receive a discharge of certain debts.

To learn more about Chapter 13, visit http://www.usdoj.gov/ust; it's the website of the U.S. Trustee Program, the organization within the U.S. Department of Justice that supervises bankruptcy cases and trustees.

If you have a mortgage through the Federal Housing Administration (FHA) or Veterans Administration (VA), you may have other foreclosure alternatives. Contact the FHA (http://www.fha.gov) or VA (http://www.homeloans.va.gov) to discuss your options. Stay in your home during the process, since you may not qualify for certain types of assistance if you move out. Renting your home will change it from a primary residence to an investment property and most likely will disqualify you for any additional "work-out" assistance from the servicer. If you choose this route, be sure the rental income is enough to help you get and keep your loan current.

Consider Giving Up Your Home without Foreclosure

Not every situation can be resolved through your loan servicer's foreclosure prevention programs. If you're not able to keep your home, or if you don't want to keep it, consider the following options:

Selling your house Your servicers might postpone foreclosure proceedings if you have a pending sales contract or if you put your home on the market. This approach works if proceeds from the sale can pay off the entire loan balance plus the expenses connected to selling the home (for example, real estate agent fees). Such a sale also would allow you to avoid late and legal fees and damage to your credit rating, and it would protect your equity in the property.

Short sale Your servicers may allow you to sell the home yourself before it forecloses on the property, agreeing to forgive any shortfall between the sale price and the mortgage balance. This approach avoids a damaging foreclosure entry on your credit report. You still may face a tax liability on the amount of debt forgiven. Consider consulting a financial adviser, accountant, or attorney for more information.

Deed in lieu of foreclosure You voluntarily transfer your property title to the servicer (with the servicer's agreement) in exchange for cancellation of the remainder of your debt. Though you lose the home, a deed in lieu of foreclosure can be less damaging to your credit than a foreclosure. You will lose any equity in the property, and you may face an income tax liability on the amount of debt forgiven. A deed in lieu may not be an option for you if other loans or obligations are secured by the property or your home.

Housing and Credit Counseling

You don't have to go through the foreclosure prevention process alone. A counselor with a housing counseling agency can assess your situation, answer your questions, go over your options, prioritize your debts, and help you prepare for discussions with your loan servicer. Housing counseling services usually are free or low cost.

While some agencies limit their counseling services to homeowners with FHA mortgages, many others offer free help to any homeowner who is having trouble making mortgage payments. Call the local office of the U.S. Department of Housing and Urban Development (http://www.hud.gov) or the housing authority in your state, city, or county for help in finding a legitimate housing counseling agency nearby. Or consider contacting the NeighborWorks' Center for Foreclosure Solutions at 1-888-995-HOPE or http://www.nw.org. The center is an initiative of NeighborWorks America.

Foreclosure Rescue Scams: Another Potential Stress for Homeowners in Distress

The possibility of losing your home to foreclosure can be terrifying. The reality that scam artists are preying on the vulnerability of desperate homeowners is equally frightening. Many so-called foreclosure rescue companies or foreclosure assistance firms claim they can help you save your home. Some are brazen enough to offer a money-back guarantee. Unfortunately, once most of these foreclosure fraudsters take your money, they leave you much the worse for wear.

Fraudulent foreclosure "rescue" professionals use half-truths and outright lies to sell services that promise relief and then fail to deliver. Their goal is to make a quick profit through fees or mortgage payments they collect from you but do not pass on to the lender. Sometimes they assume ownership of your property by deceiving you, the homeowner. Then, when it's too late to save your home, they take the property or siphon off the equity. You've lost your home to foreclosure despite your best intentions.

Foreclosure rescue firms use a variety of tactics to find homeowners in distress: Some sift through public foreclosure notices in newspapers and on the Internet or through public files at local government offices, and then send personalized letters to homeowners. Others take a broader approach through ads on the Internet, on television, or in the newspaper; posters on telephone poles, median strips, and at bus stops; or flyers or business cards at your front door. The scam artists use simple and straightforward messages, such as the following:

- "Stop Foreclosure Now!"
- "We guarantee to stop your foreclosure."
- "Keep your home. We know your home is scheduled to be sold. No problem!"
- "We have special relationships within many banks that can speed up case approvals."
- "We Can Save Your Home. Guaranteed. Free Consultation."
- "We stop foreclosures every day. Our team of professionals can stop yours this week!"

Once they have your attention, they use a variety of tactics to get your money:

- Phony counseling or phantom help.
- The scam artist tells you that he can negotiate a deal with your lender to save your house if you pay a fee first. You may be told not to contact your lender, lawyer, or credit counselor, and to let the scam artist handle all the details. Once you pay the fee, the scam artist takes off with your money.

A foreclosure prevention "specialist" really is a phony counselor who charges outrageous fees in exchange for making a few phone calls or completing some paperwork that a homeowner could easily do for himself or herself. None of the actions results in saving the home. This scam gives homeowners a false sense of hope, delays them from seeking qualified help, and exposes their personal financial information to a fraudster.

Sometimes the scam artist insists that you make all mortgage payments directly to him while he negotiates with the lender. In this instance, the scammer may collect a few months of payments before disappearing.

Bait-and-Switch You think you're signing documents for a new loan to make your existing mortgage current. This is a trick—you've signed documents that surrender the title of your house to the scam artist in exchange for a "rescue" loan.

Rent-to-Buy Scheme You're told to surrender the title as part of a deal that allows you to remain in your home as a renter, and to buy it back during the next few years. You may be told that surrendering the title will permit a borrower with a better credit rating to secure new financing—and prevent the loss of the home. But the terms of these deals usually are so burdensome that buying back your home becomes impossible. You lose the home, and the scam artist walks off with all or most of your home's equity. Worse yet, when the new borrower defaults on the loan, you're evicted.

In a variation, the scam artist raises the rent over time to the point that the former homeowner can't afford it. After missing several rent payments, the renter—the former homeowner—is evicted, leaving the "rescuer" free to sell the house.

In a similar equity-skimming situation, the scam artist offers to find a buyer for your home, but only if you sign over the deed and move out. The scam artist promises to pay you a portion of the profit when the home sells. Once you transfer the deed, the scam artist simply rents out the home and pockets the proceeds while your lender proceeds with the foreclosure. In the end, you lose your home—and you're still responsible for the unpaid mortgage. That's because transferring the deed does nothing to transfer your mortgage obligation.

Bankruptcy Foreclosure The scam artist may promise to negotiate with your lender or to get refinancing on your behalf if you pay a fee up front. Instead of contacting your lender or refinancing your loan, though, the scam artist pockets the fee and files a bankruptcy case in your name—sometimes without your knowledge.

A bankruptcy filing often stops a home foreclosure, but only temporarily. What's more, the bankruptcy process is complicated, expensive, and unforgiving. For example, if you fail to attend the first meeting with the creditors, the bankruptcy judge will dismiss the case and the foreclosure proceedings will continue.

If this happens, you could lose the money you paid to the scam artist as well as your home. Worse yet, a bankruptcy stays on your credit report for ten years and can make it difficult to obtain credit, buy a home, get life insurance, or sometimes get a job.

Where to Find Legitimate Help

If you think you may be facing foreclosure, the Federal Trade Commission (FTC), the nation's consumer protection agency, wants you to know how to recognize a foreclosure rescue scam. And even if the foreclosure process has already begun, the FTC and its law enforcement partners want you to know that legitimate options are available to help you save your home.

If you're having trouble paying your mortgage or you have gotten a foreclosure notice, contact your lender immediately. You may be able to negotiate a new repayment schedule. Remember that lenders generally don't want to foreclose; it costs them money.

Other foreclosure prevention options, including reinstatement and forbearance, are explained in *Mortgage Payments Sending You Reeling? Here's What to Do*, a publication from the FTC. Find it at http://www.ftc.gov.

You also may contact a credit counselor through the Homeownership Preservation Foundation (HPF), a nonprofit organization that operates a national 24/7 toll-free hotline (1-888-995-HOPE) with free, bilingual, personalized assistance to help at-risk homeowners avoid foreclosure. HPF is a member of the HOPE NOW Alliance of mortgage servicers, mortgage market participants, and counselors. More information about HOPE NOW is available at http://www.995hope.org.

Red Flags

If you're looking for foreclosure prevention help, avoid any business that:

- guarantees to stop the foreclosure process—no matter what your circumstances
- instructs you not to contact your lender, lawyer, or credit or housing counselor
- collects a fee before providing you with any services
- accepts payment only by cashier's check or wire transfer
- encourages you to lease your home so you can buy it back over time
- tells you to make your mortgage payments directly to it, rather than to your lender
- tells you to transfer your property deed or title to it
- offers to buy your house for cash at a fixed price that is not set by the housing market at the time of sale
- offers to fill out paperwork for you
- pressures you to sign paperwork you haven't had a chance to read thoroughly or that you don't understand

If you're having trouble paying your mortgage or you have gotten a foreclosure notice, contact your lender immediately.

Report Fraud

If you think you've been a victim of foreclosure fraud, contact:

- the Federal Trade Commission
- your state attorney general
- your Better Business Bureau

For More Information

To learn more about mortgages and other credit-related issues, visit http://www.ftc.gov/credit and MyMoney.gov, the U.S. government's portal to financial education.

The FTC works for the consumer to prevent fraudulent, deceptive, and unfair business practices in the marketplace and to provide information to help consumers spot, stop, and avoid them. To file a complaint or to get free information on consumer issues, visit http://www.ftc.gov or call toll-free 1-877-FTC-HELP (1-877-382-4357); TTY: 1-866-653-4261. The FTC enters Internet, telemarketing, identity theft, and other fraud-related complaints into Consumer Sentinel, a secure online database available to hundreds of civil and criminal law enforcement agencies in the United States and abroad.

MOVING

Being married in the military has more than its fair share of challenges, and frequent moves are one of them. No sooner have you got used to one base and met lots of new friends, you receive permanent change of station (PCS) orders and have to pack up all over again. It is even more difficult if you have children in school and family nearby.

Planning every detail of the move is the secret to success. First, treat the move as an new adventure—something to be looked forward to as a new exciting chapter in your marriage. If your partner is deployed and still receives PCS orders, you will need a power of attorney to get things done.

Start making lists—lists of bills to be paid, utilities and services to be canceled or transferred, people to be informed, important telephone numbers, and so on.

Gather documents that you will need—such as orders, medical records, birth certificates, marriage certificates, powers of attorney, living will, and insurances.

Create a countdown calendar and enter everything that needs doing three months ahead of time (if you have that luxury of planning time), everything that needs doing one month out, one week out, and on the day. Keep referring to this calendar and add to it as you think of new things that have to be done.

Check out your new base and what facilities and services are available. One of the great things about the military is that you may well have friends at the new base who can give you useful tips.

Check out the local Chamber of Commerce and Convention and Visitors Bureau if there is one—for job opportunities and to familiarize yourself with your new home area.

Stay in touch with family and friends—they will be a great support group during your move.

Your relocation office should be your first stop when you are considering a move or when you actually receive orders. This office is staffed by trained professionals who will help manage your move. Their goal is to connect you to the right resources at the right time so that you can execute an efficient and cost-effective move within the military system.

Your installation relocation office can help you:

- determine your PCS allowances
- connect with your new installation's relocation office
- create and customize a moving calendar
- connect with important installations agencies
- create a customized booklet of resources
- access a loan closet
- understand out-processing requirements
- obtain a sponsor

Permanent Change of Station (PCS) Allowances

There are various allowances associated with most moves within the continental United States (CONUS) and outside the continental United States (OCONUS). *Do not assume* that you will receive any of these allowances. Allowances change periodically. Check with the finance office on your installation to determine the exact amount of your allowances. For additional information, you can visit the Per Diem, Travel, and Transportation Allowance Committee website, the official source for the most up-to-date changes to benefits and allowances.

Housing Office

Your installation housing office can help you with:

- determining your housing allowances
- determining availability of government housing at your new location
- understanding your housing privatization options at your new location
- finding local community housing at your new location
- arranging for temporary lodging

The DoD's Automated Housing Referral Network (http://www.ahrn.com) can help you look for housing.

Personally Owned Vehicle (POV)

One POV belonging to you or your family member may be shipped at government expense overseas. It must, however, be for your or your family member's personal use only. If you desire to make your own arrangements and ship an additional POV, consult your transportation office for any restrictions that may apply. You may be required to pay an import duty on a second POV. At the option of the member shipping a vehicle overseas, a motorcycle or moped may be considered a POV if the member does not ship a vehicle with four or more wheels under the same set of military orders. A vehicle under a long-term lease (twelve months or longer) may be shipped if you obtain written permission from the leasing company.

The POV should be delivered to the port prior to the departure of the member on whose orders the shipment is to be made. This includes dependent travel authorizations when no POV has been previously shipped on the sponsor's orders. The member must have a minimum of twelve months remaining on overseas tour at the time the vehicle is delivered to the loading port. If a military spouse delivers the vehicle to the loading port, he or she must have a special power of attorney.

Personally procured transportation moves (DITY moves) allow you to personally move household goods and collect an incentive payment up to 95 percent of the government's estimate to move your household goods. You can do a personally procured move (PPM) when you have PCS orders, temporary duty assignment, separation or retirement, or assignment to, from, or between government quarters. You can use certain vehicles to move your household goods instead of having the government ship them. You may use this option to move all or a portion of their authorized JFTR weight allowance. All of the details can be found at http://www.move.mil/.

Chapter Six

Getting Married and Having a Family

When most people get married they inherit a second family of in-laws. When you marry into the military you get a very large third family. Your new third family—whether it is Army, Navy, Air Force, Marines, or Coast Guard—can also be quite a culture shock. They often speak what appears to be a different language and use expressions and acronyms that take a lot of getting used to. They tell the time differently. They have rules and standards of conduct that you as a military spouse are expected to follow, and ways of doing things that may be very strange to you at first. There is the chain of command structure, and you have to learn to recognize ranks and insignias. Living on base involves heightened levels of security. And this is all on top of getting married and setting up a home.

The good news is that while it may all seem very strange and even daunting at first, there is a lot of help on hand. Other married couples will be quick to welcome you and help you get organized. There are support groups and clubs, social workers, chaplains, and a host of other people all waiting to help you get acclimatized and familiar with your new third family.

GETTING MARRIED

Marrying into the military can be a daunting task, but tens of thousands do it successfully every year and have stunningly happy marriages. Others perhaps rush in without considering all the consequences, and it is these marriages that often face challenges down the road.

If you are marrying someone who is serving in the military, especially at a time when we are fighting a vicious enemy overseas on several fronts, there are several issues that have to be addressed that a civilian couple doesn't even think about.

The most important question to ask is "Can I handle the demands of married life in the military?"

When you marry you will probably have to move away from home—away from your family and friends—to live on or close to the base where your partner is stationed. That is part of the commitment you make to the person you love. However, you will still be able to visit family and friends, and they can visit you—and within a short time you will have made lots of new friends as well.

The chances are that your partner will be deployed, and this can mean long and lonely separations. Deployment also brings with it the risk of injury or worse. You will worry all the time your partner is away, but you will still be able to communicate with each other, and there are strong support groups available if things start to get too much. Many couples say that deployment made their marriages stronger because it helped them focus on what was really important and precious in their lives. As one Army wife said, "Love isn't enough to make a marriage succeed. It also takes commitment and sacrifice. Your marriage will only be as successful as you want it to be."

Apart from the military considerations, there may be ethnic and cultural issues as well that have to be thought about. You and your partner may have different religions with different holidays and traditions. If you have children, what religion will they be brought up in? What holidays will you celebrate? You and your partner may have been born in different countries and have different mother tongues. In some societies younger relatives are expected to care for elderly members of the family when they become sick or can't look after themselves. What would you do if your partner asked that an elderly aunt move in with you?

These sorts of issues should not be an impediment to getting married, but they have to be discussed openly. Traditions that might seem unimportant to you might be extremely important to your partner. Talk about them and other cultural traditions. Respect each other's position and point of view. Remember that talking problems over usually helps them go away.

There are lots of other things you can do to cement your relationship. You can each learn about your partner's history and cultural background through reading books or joining societies. If you were brought up speaking different languages, make a commitment to each other to become fluent in both. It will be a bond between you, and it will be invaluable when you both meet your partner's friends and family.

It is also important to talk about your plans to get married with your parents. If you are marrying someone in the military, your parents will probably be very concerned. No matter how supportive they are, they will be worried about what might happen if your partner is deployed and what might happen to you.

Talk it over with your parents, introduce your partner to them, and demonstrate through your love that this is the right thing to do.

If you're in the military and stationed in the United States, getting married as a member of the military is much the same as a civilian marriage. You don't need advance permission, and there is no special military paperwork to fill out before the marriage. You simply obtain a marriage license off-base and get married according to the laws of the state where the marriage is taking place.

If you are overseas and marrying a foreign national, it's a different story. There are many forms to complete; you must obtain counseling and your commander's permission (which is rarely withheld without very good reason); your spouse must undergo a security background check and pass a medical examination. Finally, the marriage has to be recognized by the U.S. embassy. The entire process can take several months.

Regardless of where or who, once married, if the spouse is nonmilitary, the military member can bring a copy of the certified marriage certificate to the personnel headquarters on the base to receive a dependent ID card for the spouse and enroll the spouse in the Defense Eligibility Enrollment Reporting System (DEERS), in order to qualify for military benefits such as medical coverage and commissary and base exchange privileges.

Timing can be important in a military marriage. If you have permanent change of station (PCS) orders and get married before you actually make the move, you can have your spouse added to your orders and the military will pay for the relocation of your spouse and his or her property (furniture and such). However, if you report to your new duty assignment first and then get married, you will have to pay for the relocation of your spouse out of your own pocket.

Actually making the move means reporting into your new base. So you can leave your old base, take leave (vacation), get married, report in to your new base, get your orders amended to include your new spouse, and the military will pay for the spouse's move. However, if you report to your new base and then take leave to get married, you're on your own when it comes to moving expenses for your spouse.

If you want to get married on base, the point of contact is the chaplain's office. Each military base has one (or more) chapels that are used for religious services. One can get married in a base chapel, just as one can get married in a church off-base. Base chaplains offer a complete variety of marriage choices, including religious (almost any denomination), nonreligious, casual, civilian-formal, and military-formal.

If the wedding is conducted by a military chaplain, there is never a fee. By regulation, chaplains cannot directly accept donations. One can make a donation to the chaplain's fund, however, during a normal worship service.

Military Formal Weddings

The military formal wedding would entail the following: An officer or enlisted personnel in the bridal party wear uniforms in accordance with the formality of the wedding and seasonal uniform regulations. For commissioned officers, evening dress uniform is the same as civilian white tie and tails. The dinner or mess dress uniform is equivalent to civilian "black tie" requirements. The choice to attend the wedding in uniform as a military guest is optional.

In the case of noncommissioned officers and other enlisted, dress blues or Army green uniforms may be worn at formal or informal weddings. A female military member (officer or enlisted) may be married in uniform, or she may wear a traditional bridal gown. A boutonniere is never worn with a uniform.

The "arch of sabers" is usually part of a military formal wedding. The arch of swords takes place immediately following the ceremony, preferably when the couple leaves the chapel or church, on the steps or walk. Since a church is a sanctuary, in case of bad weather, and with permission, the arch may be formed inside the chapel or church. Also, with permission, you may be allowed to have two arches of sabers, one in the church and one outside. White gloves are a necessity for all saber (sword) bearers.

Marrying Another Military Member

If you are marrying someone in the military rather than a civilian, there's one primary difference, and that's in the area of housing benefits allowed after the marriage, rather than actual marriage procedures.

There are two basic types of housing allowance (monetary allowance paid to military members who live off-base): single allowance and "with dependent" allowance. Usually, single (nonmarried) military members who are allowed to live off-base receive the single allowance. Those who have dependents (civilian spouse and/or children) receive a larger allowance called the "with dependent" allowance.

If two military members marry (assuming there are no children), each receives the single allowance. The total of both of these single allowances is always more than the "with dependent" allowance. For example, a military member in the rank of E-4, stationed at Fort McClellan, Alabama, who married a civilian, would receive $525 per month for a housing allowance. If a military member married another military member, they would *each* receive the single rate, which would be $424 per month.

If a military member marries another military member and they have children, one member will receive the "with dependent" rate, and the other member will receive the "single" rate. Usually, the member with the most rank receives the "with dependent" rate, because it means more money each month.

Being Posted Together

Each of the services has a program called "Join-Spouse," in which the services try as hard as they can to station spouses together, or at least within one hundred miles of each other. However, there is absolutely no guarantee. In order for spouses to be stationed together, there have to be "slots" (job positions) available to assign them to.

For example, let's say that an Air Force B-1 aircraft mechanic married a Navy F-14 aircraft mechanic. Because the B-1 bomber is only stationed at certain Air Force bases, and because the F-14 Tomcat Fighter Aircraft is only stationed at certain Navy bases, this couple is probably never going to be stationed together. The best the services could do would be to try to find a B-1 base as close as possible to an F-14 base (and, if this case, that would be at least one thousand miles away).

If a military person marries a person in the same service, the chances of getting stationed together are better. Each of the services brag about an 85 percent success rate with in-service Join-Spouse. (That sounds pretty good until you realize that fifteen out of one hundred military couples in each service are not stationed together.)

When one marries someone in a different service, it becomes more complicated. The success rate of "Join-Spouse" goes down dramatically to somewhere around 50 percent.

THE HONEYMOON

Most military members get two weeks or more notice before actually leaving on a deployment. So it's possible that a commander would grant a couple of days' leave during that two weeks—but not much leave, as there is much to do before a unit can deploy.

Otherwise, there are two primary ways to get time off in the military. The first is called a "pass," which is basically normal time off (like holidays and weekends) and special time off that might be granted by a commander or supervisor (up to seventy-two hours). The second way is leave (vacation time). Every military member gets thirty days of leave per year, earned at the rate of 2.5 days per month.

A commander and/or supervisor could grant a pass (up to three days) for a member to get married and/or honeymoon, or the military member could take up to thirty days' leave (assuming he or she had that much leave saved up, and the unit could afford to lose him or her for that long a period).

Make the most of your honeymoon time. Spend as little of it traveling as possible; save the far-off journeys for when you can spare the days. Also, if you can afford to, splurge on one night in a great hotel rather than two in a mediocre one. And let the world (or at least the reservations clerk) know that you're on your honeymoon. It's true that all the world loves a lover, and you never know what goodies or upgrades may come to you gratis.

Whether on a honeymoon or not, military members can travel "space available" for free on military aircraft to locations around the world. If available leave time is a factor, Space-A travel might not be viable. To travel Space-A, a military member must already be on leave. Sometimes it can take several days for a flight with space available on it to be going in your direction. Also, one wants to make sure he or she has adequate funds to buy a return ticket, in case the passenger can't find a Space-A flight available going back to the originating base.

Check out the Armed Forces Vacation Club. This program allows military members to rent luxury condos around the world for $249 per week. In addition, many hotels and resorts offer military discounts; it always pays to ask. If money is an issue, military couples could stay in billeting on any military base for about $16 to $20 per night—if you don't mind spending your honeymoon on a military base.

Consider the tradition of giving a wedding night gift to your new spouse. It doesn't have to be large or expensive (you can even make it yourself), just some object that can serve as a sentimental reminder of your first night together as a married couple. Then, if orders come through and you're separated for a time, you'll have something wonderful to hold onto until you're reunited.

If you are planning on joining the military and planning on getting married, there are certain advantages (as well as some disadvantages) to tying that knot before you leave for basic training.

However, we emphasize that one should not make their marriage decision based primarily on these factors. The divorce rate in the United States is about 50 percent, and that statistic follows over to the military. In fact, the military divorce rate may be even a little higher, because of the difficulty of a military life (frequent moves, unaccompanied assignments, long working hours, combat deployments, etc.).

But if you've already made your decision to get married and are now just deciding whether it would be better to get married before or after joining the military, the following information may be of use to you.

Housing Allowance

A married service member receives a housing allowance while in basic training and follow-on job training (technical school, AIT, A-school), in order to provide a household for his or her dependents, even though they are also living for free in government quarters (barracks). If you get married before joining the military, this tax-free housing allowance begins on the very first day of active duty (the first day of basic training).

If one waits until after joining the military to get married, the housing allowance becomes effective on the date of the marriage. However, one needs a certified copy of the marriage certificate to change marital status, and (depending on the state) this can take a couple of weeks or even a month to obtain. Even so, the housing allowance would be backdated to the date of the marriage.

Medical Care

Dependents of active-duty members are covered by the Military Medical System (TRICARE), effective the very first day of active duty. During basic training in-processing, the recruit completes paperwork to enroll his or her dependents in DEERS and for a military dependent ID card. The ID card paperwork is mailed to the spouse, who can then take it to any military installation and obtain a military dependent ID card. If medical care is needed before getting the ID card, the spouse can keep the medical receipts and file for reimbursement later, under the TRICARE Standard or TRICARE Extra program (depending on whether or not the medical provider is part of the TRICARE network).

Family Separation Allowance

Married members are entitled to a family separation allowance when they are separated from their dependents due to military orders. The tax-free allowance begins after separation of thirty days. This means married people in basic training and technical school (if the technical school duration is less than twenty weeks) begin to receive this pay thirty days after going on active duty. Single personnel do not receive this allowance.

Movement of Dependents

Unless the first duty assignment is an unaccompanied (remote) overseas tour, the married military member is entitled to move their dependents (and personal property) to the first duty station at government expense. Travel entitlements end when one signs in at the new duty station, so whether or not one can be reimbursed for dependent travel depends on the date of the marriage.

For example, Airman Jones graduates technical school (Air Force job training), then goes home on leave en route to his first duty assignment. While on leave, Airman Jones gets married. He then reports to his first duty station. He will be entitled to movement of dependents at government expense, because the date of the marriage was before he signed in at the duty station.

Another example: PFC Jackson finishes AIT (Army job training) and goes to his first duty assignment. A couple of weeks later, his fiancée flies down, and they get married. PFC Jackson cannot move his wife and her property to the duty assignment at government expense, because the marriage occurred after he completed his assignment move.

There is an exception to the above rule for certain overseas assignments. When a single person is assigned to a "long" overseas tour, the assignment length is generally twenty-four months (the unaccompanied tour length). For an accompanied married person, the tour length is usually thirty-six months. If a single person goes overseas on such a tour and gets married during the tour, he or she can apply to move his dependents overseas, if he or she agree to extend the tour length to the accompanied tour length.

In order to move dependents at government expense, one's orders must include authorization to do so. This means that if one gets married at the last minute before leaving job training, or if one gets married on leave en route to the first assignment, the orders won't have this authorization on it, and they will have to be amended after arrival. As the military rarely does paperwork very fast, this amendment process can sometimes take several weeks. This will delay the reimbursement of dependent moving expenses. If the member had gotten married before joining the military, his or her orders would have had dependent moving entitlements annotated originally and would thereby avoid this delay in reimbursement.

Job Training

If technical school, AIT, or A-school is twenty weeks or longer in duration (at a single location), one is entitled to move one's dependents to the school location at government expense. He or she is then (usually thirty days after arrival) allowed to live with the dependents after duty hours. Single members, of course, cannot move their girlfriend or boyfriend at government expense, nor will they be allowed to live off-base (even at their own expense) at job training locations.

In such cases, if the military member elects not to move his or her dependents, the family separation allowance stops, because the member is not being forced to be separated (the dependents are allowed to move at government expense, so if they don't move, that's the member's choice). Of course, if the dependents do join the member, family separation allowance stops as well, as the member is no longer separated from his or her dependents due to military orders.

If the job training is less than twenty weeks, a married person can still elect to move his or her dependents (at his or her own expense) and would (usually) be allowed to live with them off-base (beginning thirty days after arrival), with the school commander's permission (as long as the student is doing okay in class, such permission is routinely granted). If the dependents do move to the member's school location, family separation allowance stops.

Qualifications

The services require additional paperwork, additional processing, and sometime even waivers for members with dependents. For example, the Air Force requires a credit check for any member who is married or has ever been married. If you're in the delayed enlistment program (DEP) and decide to get married before shipping out to basic training, you'll want to check with your recruiter to determine (depending on what additional processing is required and when you're shipping out) whether this would possibly delay your shipping date.

Dependent Support

All of the services have regulations that require military members to provide adequate support to their dependents. In fact, while in basic training and job school, you're being provided a housing allowance for the sole purpose of providing a place to live for your dependent family members. If your spouse makes an official complaint to your commander that you are failing or refusing to provide financial support, you could be in a heap of trouble.

While one doesn't want to think about divorce when one is anxious to get married, divorce is a real possibility (remember the 50 percent statistic). Military members should be aware that there is a special law that applies to them when it comes to divorce and retirement pay. The Uniformed Services Former Spouse Protection Act allows any state court to treat your *future* military retired pay as joint property, to be divided with your spouse, in the event of a divorce. (Source: Rod Powers, at http://usmilitary.about.com.)

THE NEWLYWEDS

The honeymoon may be over, but now is the time to get down to the serious business of organizing your new lives together. There are all sorts of procedures to be followed, benefits to sign up for, and special deals to take advantage of. There is health care, great shopping at the Exchange stores and commissaries, and a wide range of facilities and activities to get involved with—from bowling and movie theaters to fitness centers and social clubs.

If you haven't done it already, you must notify your command that you are married and have a spouse. If you are the serving member, you become your spouse's sponsor, and as such, you have to fill in a lot of paperwork on his or her behalf.

One of your next stops should be to the family support center (FMS). It will have all the latest information and knowledgeable people ready to help you acclimatize. Many bases run orientations for newcomers, especially for spouses. The FMS will also be able to tell you where to go to sign up for benefits, classes, orientations, and so on.

Family Support Centers

Each branch of the military has its own name for family support centers and their own programs for new spouses. These are as follows:

- Army Community Service Center; Family Team Building
- Airman and Family Readiness Center; Heartlink
- Marine Corps Community Services Center: LINKS (Lifestyle, Insights, Networking, Knowledge, and Skills)
- Navy Fleet and Family Support Center: COMPASS

However, before you can start doing the paperwork, you will have to assemble a number of important documents. You will need the following:

- An original copy of your marriage certificate—and make several copies of it.
- A DEERS enrollment form (see below). It is up to you to enroll—it is not automatically done for you.
- Birth certificates for your spouse and all children.
- Social Security cards for your spouse and all children.
- Photo ID.

ID Cards

All spouses and children age ten and over must have a military ID card. You apply for your ID card using Department of Defense Form 1172, and there is an accompanying booklet that explains how to fill it in. You will also have to produce birth certificates, your marriage license, and a photo ID.

You will need your ID card to get on base and when using commissaries and the like. You will also have to show it when getting medical treatment (see below).

Vehicle Registration

Remember to register your vehicle and get a sticker that will allow you to enter the base or installation. You normally register through the Military Police, and it is worth a call to them beforehand to see what documents you need to bring—usually your driver's license, proof of insurance, and registration document.

Legal Considerations

While it may be the last thing you want to think about, you should establish power of attorney for your enlisted spouse, especially if he or she is likely to be deployed. This will allow you to make decisions on his or her behalf if your spouse is unable to, and it is important that you talk through various scenarios so that you are absolutely sure you will be doing what your spouse wants. Again, most bases have a legal assistance office that can help you with this.

Legal Assistance

You and your dependents are eligible to receive military legal assistance. You can get most of the legal services you could get from a civilian attorney, except appearance in a civilian court. Among the available services are the following:

- Advice and assistance with personal problems of a civil nature, such as marriage, divorce, adoption, civil damage actions, insurance, indebtedness, and contracts
- Advice on tax matters and forms
- Preparation of wills and powers of attorney
- Notary public services
- Advice concerning sale or lease of real property
- Assistance in obtaining applications for certificates of citizenship and naturalization

Medical

All military spouses are entitled to medical benefits through DEERS. You enroll at the uniformed services personnel office, and you can find the one nearest to you at http://www.militaryinstallations.dod.mil.

DEERS Enrollment

Proper registration in the DEERS is key to receiving timely and effective medical benefits. DEERS is a worldwide, computerized database of uniformed services members (sponsors), their family members, and others who are eligible for military benefits, including TRICARE.

All sponsors (active-duty, retired, National Guard, or Reserve) are automatically registered in DEERS. However, the sponsor must register eligible family members. After family members are registered, they can update personal information such as addresses and phone numbers.

To use TRICARE benefits, you must have a valid uniformed services or military ID card, which you can obtain from your nearest ID card office, and you must be listed in the DEERS database. The ID card states on the back, in the "medical" block, whether you are eligible for medical care from military or civilian sources.

When getting care, your provider will ask to see a copy of your ID card and will make copies for his or her records. Please be sure to have your ID card with you whenever you are getting care or having prescriptions filled.

Children

Children under age ten can generally use a parent's or guardian's ID card, but they must be registered in DEERS. At age ten, the sponsor must obtain an ID card for the child. Children under age ten should have an ID card of their own when in the custody of a parent or guardian who is not eligible for TRICARE benefits or who is not the custodial parent after a divorce. If both parents are active-duty service members, then either may be listed as the child's sponsor in DEERS.

Stepchildren and adopted children can also be enrolled, provided they live with the service member and spouse and are not already the dependent of another service member. You will need birth certificates and final adoption decree in the case of an adopted child. If the child is from a former marriage, you will also need a copy of the divorce decree or a death certificate if your former spouse is deceased and any custody documents.

Health Benefits

You may want to change to a family health benefits enrollment. You may enroll or change enrollment from "Self Only" to "Self and Family," from one plan or option to another, or make any combination of these changes during the period beginning thirty-one days before and ending sixty days after a change in your family status. Otherwise, you will have to wait until the next health benefits open season to make the change. Review your pay withholding information and, if necessary, change it.

If you want to provide immediate coverage for your new spouse, you may submit an enrollment request during the pay period before the anticipated date of your marriage. If the effective date of the change is before your marriage, your new spouse does not become eligible for coverage until the actual day of your marriage.

If you enroll or change your enrollment before the date of your marriage and intend to change your name, you must note on your request: "Now: [Current Name] will be: [Married Name]." You must also give the reason for the change and the date of the marriage in your request.

STARTING A FAMILY

If you are thinking of starting a family, ensure that your health insurance covers you as the mother throughout your pregnancy for prenatal care and delivery and then covers you and your child thereafter. Consult with your health benefits adviser.

You should start taking care of yourself *before* you start trying to get pregnant. This is called preconception health. It means knowing how health conditions and risk factors could affect you or your unborn baby if you become pregnant. For example, some foods, habits, and medicines can harm your baby—even before he or she is conceived. Some health problems also can affect pregnancy.

Talk to your doctor before pregnancy to learn what you can do to prepare your body. Women should prepare for pregnancy before becoming sexually active. Ideally, women should give themselves at least three months to prepare before getting pregnant.

The five most important things you can do before becoming pregnant are the following:

1. 1. Take 400 micrograms (400 mcg or 0.4 mg) of folic acid every day for at least three months before getting pregnant to lower the risk of some birth defects of the brain and spine. You can get folic acid from some foods. But it's hard to get all the folic acid you need from foods alone. Taking a vitamin with folic acid is the best and easiest way to be sure you're getting enough.
2. Stop smoking and drinking alcohol. Ask your doctor for help.
3. If you have a medical condition, be sure it is under control. Some conditions include asthma, diabetes, depression, high blood pressure, obesity, thyroid disease, or epilepsy. Be sure your vaccinations are up to date.
4. Talk to your doctor about any over-the-counter and prescription medicines you are using. These include dietary or herbal supplements. Some medicines are not safe during pregnancy. At the same time, stopping medicines you need also can be harmful.

5. Avoid contact with toxic substances or materials at work and at home that could be harmful. Stay away from chemicals and cat or rodent feces.

Having a baby means lots of changes to your lives and lifestyle. You will probably already have converted a bedroom into a nursery and bought all the things your new child will need. And it probably came as a shock just how much everything cost—and that's just the beginning!

According to the U.S. Department of Agriculture's report "Expenditures on Children by Families," a couple earning between $56,870 and $98,470 will spend $221,190 on a child up until age seventeen. That is why it is so important to have your finances in order. As soon-to-be parents, you need to take a hard look at your financial situation and start to plan for the future.

Conduct a financial audit—what are your monthly outgoings, what other financial commitments do you have, what is your credit card debt, and so on? If one of you will have to give up work, how will that impact your finances? Once you know exactly what your financial situation is, you can decide if cuts need to be made in some areas to free up money to pay for the baby's upkeep. In the short term you may have to find extra money to pay for child care, and in the long term you may want to put money aside for your child's college education. Even a small amount put aside every week or month will mount up over the years, and you can let family and friends know in case they want to make a gift in lieu of birthday or Christmas presents.

Check health and life insurance to ensure you have adequate coverage. While most young couples don't think about life insurance, it is exactly the right time to take it out because the younger you are the lower the premiums you pay. While the serving member of the marriage will have some coverage, it is a good idea for both partners to have life and disability coverage. Make sure you apply for and receive all the benefits and leave you are entitled to.

As soon as possible after the delivery, get a Social Security number for your baby, and update your will to include your new child as a beneficiary if that is your wish.

You will find that family and friends will buy lots of clothes and toys for the new baby, which will save you money, but other costs, such as diapers, a crib, a baby seat for the car, and so on, will have to be met.

Young mothers can join a young mothers club and start swapping baby clothes and other items. Babies grow so fast they don't wear their clothes out, so why not pass them on? You can also find great bargains at garage sales, thrift shops, and the like.

Self and Family

A Self and Family enrollment provides benefits for you and your eligible family members. All of your eligible family members are automatically covered, even if you didn't list them on your "Health Benefits Election Form" (SF 2809) or another appropriate request. You cannot exclude any eligible family member, and you cannot provide coverage for anyone who is not an eligible family member.

You may enroll for Self and Family coverage before you have any eligible family members. Then a new eligible family member (such as a newborn child or a new spouse) will be automatically covered by your family enrollment from the date he or she becomes a family member. When a new family member is added to your existing Self and Family enrollment, you do not have to complete a new SF 2809 or other appropriate request, but your carrier may ask you for information about your new family member. You will send the requested information directly to the carrier. Exception: If you want to add a foster child to your coverage, you must provide eligibility information to your employing office.

If both you and your spouse are eligible to enroll, one of you may enroll for Self and Family to cover your entire family. If you have no eligible children to cover, each of you may enroll for Self Only in the same or different plans. Generally, you will pay lower premiums for two Self Only enrollments.

TRICARE Programs

TRICARE Prime is a managed-care option, similar to a civilian health maintenance organization (HMO).

Prime is for active duty service members and available to other TRICARE beneficiaries. Active duty service members (ADSM) are required to be enrolled in Prime; they must take action to enroll, by filling out the appropriate enrollment form and submitting it to their regional contractor. There is no cost to the service member.

Other TRICARE beneficiaries may be eligible for Prime. Eligibility for any kind of TRICARE coverage is determined by the uniformed services. TRICARE manages the military health care program, but the services decide who is or is not eligible to receive TRICARE coverage.

Prime enrollees receive most of their health care at a military treatment facility (MTF), and their care is coordinated by a primary care manager (PCM). Prime is not available everywhere.

Prime enrollees must follow some well-defined rules and procedures, such as seeking care first from the MTF. For specialty care, the Prime enrollee must receive a referral from his or her PCM and authorization from the regional contractor. Failure to do so could result in costly point of service (POS) option charges. Emergency care is not subject to POS charges.

TRICARE Prime Remote is the program for service members and their families who are on remote assignment, typically fifty miles from a MTF.

The TRICARE Overseas Program delivers the Prime benefit to ADSMs and their families in the three overseas areas: Europe, the Pacific, and Latin America/Canada. The TRICARE Global Remote program delivers the Prime benefit to ADSMs and families stationed in designated "remote" locations overseas.

TRICARE Standard is the basic TRICARE health care program, offering comprehensive health care coverage, for beneficiaries (not to include active duty members) not enrolled in TRICARE Prime. Standard does not require enrollment.

Standard is a fee-for-service plan that gives beneficiaries the option to see any TRICARE-certified/authorized provider (doctor, nurse practitioner, lab, clinic, etc.). Standard offers the greatest flexibility in choosing a provider, but it will also involve greater out-of-pocket expenses for you, the patient. You also may be required to file your own claims.

Standard requires that you satisfy a yearly deductible before TRICARE cost sharing begins, and you will be required to pay copayments or cost shares for outpatient care, medications, and inpatient care.

TRICARE Extra can be used by any TRICARE-eligible beneficiary, who is not active duty, not otherwise enrolled in Prime, and not eligible for TRICARE for Life (TFL).

TRICARE Extra goes into effect whenever a Standard beneficiary chooses to make an appointment with a TRICARE network provider. Extra, like Standard, requires no enrollment and involves no enrollment fee.

TRICARE Extra is essentially an option for TRICARE Standard beneficiaries who want to save on out-of-pocket expenses by making an appointment with a TRICARE Prime network provider (doctor, nurse practitioner, lab, etc.). The appointment with the in-network provider will cost 5 percent less than it would with a doctor who is a TRICARE authorized or participating provider.

Also, the TRICARE Extra option user can expect that the network provider will file all claims forms for him or her. The Standard beneficiary might have claims filed for him or her, but the nonnetwork provider can decide to file on his or her behalf or not, on a case by case basis.

Under TRICARE Extra, because there is no enrollment, there is no Extra identification card. Your valid uniformed services ID card serves as proof of your eligibility to receive health care coverage from any TRICARE Prime provider.

TRICARE For Life (TFL) is a Medicare wraparound coverage available to Medicare-entitled uniformed service retirees, including retired Guard members and Reservists, Medicare-entitled family members and widows/widowers (dependent parents and parents-in-law are excluded), Medicare-entitled Congressional Medal of Honor recipients and their family members, and certain Medicare-entitled unremarried former spouses.

To take advantage of TFL, you and your eligible family members' personal information and Medicare Part B status must be up-to-date in the Defense Enrollment Eligibility Reporting System (DEERS). You may update your information by phone (1-800-538-9552) or by visiting your nearest ID card–issuing facility. Visit www.dmdc.osd.mil/rsl to locate the nearest ID card facility.

Family Separation Allowance

Family separation allowance (currently $250 per month) is normally paid anytime a military member is separated from his or her dependents for longer than thirty days, due to military orders. For example, members with dependents attending basic training and job training (if the job training is less than twenty weeks and dependents are not authorized to relocate to the training base) receive $250 per month, beginning thirty days after separation.

The same applies to military-married-to-military, except:

1. The members must be residing together immediately prior to the departure.
2. Only one member can receive the allowance. Payment shall be made to the member whose orders resulted in the separation. If both members receive orders requiring departure on the same day, then payment will go to the senior member.

Care of Children (Dependents)

Military couples with children must develop a "family care plan" that details exactly what the care arrangements are in the event that both members must deploy. Failure to develop and maintain a workable family care plan can result in discharge.

Note: If you are a wounded warrior, there are lots of ways of getting housing grants and assistance for yourself and your family. There are grants for remodeling homes to make them disability-friendly. Habitat for Humanity has an extensive building program for combat-disabled veterans and their families.

Chapter Seven

Education—Funding Yourself and Offspring

A vast number of education programs and services are available for Armed Forces personnel, veterans, and their family members. Most individuals may even be eligible for more than one educational benefit.

Members of the Armed Forces become eligible for GI Bill benefits based on the date of entry into the service and length of time served. The Department of Veterans Affairs (VA) provides several education programs, including the Montgomery GI Bill for Active Duty and Veterans (MGIB-AD), Montgomery GI Bill for Selected Reserves (MGIB-SR), Post 9/11 GI Bill, Reserve Education Assistance Program (REAP), Veterans Education Assistance Program (VEAP), Spouse and Dependents Education Assistance (DEA), and the Vocational Rehabilitation and Education (VR&E) program.

An honorable discharge must be received to maintain eligibility for VA education benefits. It is also important to ensure the education program in which you plan to enroll has been *approved for reimbursement* under the GI Bill program for which you have established eligibility. For those utilizing Post 9/11 GI Bill benefits, your school may have entered into a partnership with the VA under the Yellow Ribbon Program to help defray tuition and fees that exceed the approved reimbursement amount established by the VA.

Each branch of the Armed Forces provides Tuition Assistance (TA) funding for active-duty military, Guards, and Reservists pursuing off-duty or voluntary education. Eligibility criteria, application procedures, and program guidelines are determined by each service branch. Military personnel may receive up to $4,500 in TA funds per fiscal year with limits of $250 per semester credit hour, $166.67 per quarter credit hour, and $16.67 per clock hour for tuition and fees for courses offered by learning institutions that are regionally or nationally accredited. TA funding may be requested through Military Education Centers.

The Tuition Assistance Top-Up Program allows military personnel to request reimbursement from the VA for the difference between the total course cost and the cost covered by TA funding, up to the maximum rate payable to a veteran. Top-Up reimbursements are deducted from available GI Bill benefits.

As part of the Higher Education Act, Congress approved the In-State Tuition Directive in 2009, which guarantees military personnel, spouses, and their college-age children in-state tuition at public colleges and universities in the state where they reside or are permanently stationed. Many states also offer *additional tuition incentives* such as grants and scholarships for military personnel and their family members.

Most Armed Forces branches provide the College Fund program, also known as a "Kicker." The College Fund, which is offered upon enlistment and based on service branch criteria, increases the monetary value of your GI Bill benefits. The Student Loan Repayment Program (LRP) is offered by military branches to partially or fully repay college loans. The loan must be in good standing (not in default) to qualify for the LRP. Eligibility is determined by your branch of service, military occupational specialties, and the terms of your enlistment contract.

Branch-specific education programs, such as eArmyU, the Marine Corps Institute, and the Community College of the Air Force, enable service members to participate in courses to earn college credits, degrees, and certificates. These programs vary by branch, may be provided at no cost or covered by tuition assistance funding, and may include books, study materials, laptop computer, printer, e-mail account, and Internet access. Commissioning programs provide opportunities for active duty, enlisted personnel to receive a scholarship to attend college to earn their degree and a commission as an officer. While attending college, service members maintain their benefits, pay, and privileges.

Navy personnel may take advantage of the Advanced Education Voucher Program or the Graduate Education Voucher Program, which provide advanced education opportunities and funding for senior enlisted personnel (E-7 through E-9) and officers to pursue Navy-relevant degree programs. Education programs have also been created for those who are deployed or on sea-duty assignments and include the Navy College-at-Sea (NCPACE), Marine Deployed Education, and SOCCOAST Afloat.

Military transcripts endorsing and recommending college credit for military education and training and recognized by the American Council on Education (ACE) are available for all members of the Armed Forces. Schools can evaluate official military transcripts to determine if college credits can be awarded based on your military training and experience. Available military transcripts include Verification of Military Education and Training (VMET), https://www.dmdc.osd.mil/appj/vmet/index.jsp; Army/American Council on Education Registry Transcript System (AARTS), http://aarts.army.mil; Sailor/Marine American Council on Education Registry Transcript (SMART), https://smart.navy.mil; Community College of the Air Force (CCAF) transcript, http://www.au.af.mil/au/ccaf/transcripts.asp; and Coast Guard transcript, http://www.uscg.mil/hq/cgi/forms/CG_Form_1561.pdf.

Military Education Centers, sponsored by all branches of the Armed Forces, provide counselors to explain education programs, use of benefits, and provide guidance; apprenticeship programs for earning MOS-related, nationally recognized certifications; testing services for Armed Forces tests, government-funded Defense Activity for Non-Traditional Education Support (DANTES) examinations (such as the SAT, ACT, GED, CLEP, DSST, and ECE), and proctoring college and home school examinations; assessment tools to analyze your personality, interests, and skills; and tuition assistance funding approval.

Servicemembers Opportunity Colleges (SOC) is a consortium of nineteen hundred appropriately accredited colleges and universities providing degree programs for service members and their adult dependent family members. These schools have flexible policies and are committed to reasonable transfer of credits and reduced residency requirements; they award academic credit for military training and experience, and credit for at least one nationally recognized testing program such as DANTES Subject Standardized Tests (DSST), College-Level Examination Program (CLEP), and Excelsior College Examinations (ECE).

The SOC Degree Network System is a subgroup of SOC Consortium member institutions selected by the military services to deliver specific associate and bachelor's degree programs to service members and their families. Through the SOCAD (Army), SOCNAV (Navy), SOCMAR (Marine Corps), and SOCCOAST (Coast Guard) programs, schools belonging to this subgroup accept credits from other network schools.

Service members and their dependents may also utilize government-funded education assistance programs, including the Tutorial Assistance program for those receiving VA benefits at a half-time or more rate; the Tutor.com program, which provides free 24/7 online tutoring and homework help from live, professional tutors; and the Online Academic Skills Course for military personnel to build their vocabulary, reading comprehension, and math skills.

Family Member Educational Assistance Programs provide grants, scholarships, no-interest loans, and tuition aid to military spouses and children. The Spouse Education Assistance Program (EAP) is available for spouses of Army service members assigned in Europe, Korea, Japan, or Okinawa. The Navy Marine Corps Relief Society (NMCRS) offers the Spouse Tuition Aid

Program (STAP), need-based, no-interest loans to spouses of active-duty Navy and Marine Corps personnel stationed in an overseas location. The My Career Advancement Account (MyCAA) program provides up to $4,000 of financial assistance (over two years) for military spouses of active-duty service members (in pay grades E1–E5, W1–W2, and O1–O2) who are pursuing associate degree programs, licenses, or credentials leading to employment in portable career fields.

The Military Community Academic Explorer (AeX) provides an unbiased, neutral platform containing information on over four thousand accredited academic institutions. AeX offers search criteria such as degree/discipline, degree level, location, and school name that provide access to information on schools that specifically match your preferences.

The Military Community Scholarship and Financial Aid Explorer (SFeX) lists over twenty-five hundred scholarships and fellowships and allows you to search specifically for opportunities for active-duty military, Guards, Reservists, veterans, spouses, high school seniors, and degree programs. SFeX includes many of the grant and scholarship opportunities offered by the service branches.

EDUCATION/TRAINING

Service members leaving the military sometimes find they need specific education or training to enter the civilian careers they want. The following section will help you identify the resources to assist you in getting the training and education you need.

Your Education Benefits: Montgomery GI Bill, Post 9/11 GI Bill, VEAP, and More

Several programs administered by the Department of Veterans Affairs (VA) provide financial assistance to veterans for education programs. This includes enrollment in degree programs, technical and vocational programs, correspondence courses, flight training courses, and on-the-job training and apprenticeship programs. To be eligible, programs must be approved, usually by a state-approving agency, for VA purposes, before VA education program benefits are paid.

Two of these programs are the Post-Vietnam-Era Veterans' Educational Assistance Program (VEAP) and the Montgomery GI Bill (MGIB). Both programs are intended to help you develop skills that will enhance your opportunities for employment. As a rule, the benefits under either of these programs must be used within ten years of separation from active duty.

Post-Vietnam-Era Veterans' Education Assistance Program (VEAP) Eligibility

With the exception of some people who signed delayed entry contracts before January 1, 1977, VEAP is for people who first entered active duty during the period January 1, 1977, through June 30, 1985, and who made a contribution to a VEAP account before April 1, 1987. If you participated in VEAP and withdrew your contribution, you may start a new allotment, or make a lump-sum contribution, at any time while you are on active duty.

Montgomery GI Bill (MGIB) Eligibility

MGIB eligibility is straightforward for most veterans, but it can be complex for others. If you have questions about MGIB eligibility, check with your Education Center or call the VA toll-free education number, 1-888-GI Bill-1 (1-888-442-4551). You may also get information at the VA Education Service website http://www.gibill.va.gov/.

With the exception of some officers who received a commission, after December 31, 1976, as a result of graduating from a service academy, or after completing a Reserve Officer Training Corps (ROTC) scholarship program, the MGIB is for people who first came on active duty on July 1, 1985, or later and who did not decline—in writing—to participate in the MGIB program.

To be eligible for the full thirty-six months of MGIB benefits, veterans must normally meet the character of service and minimum length of service requirements. Some veterans who are separated from active duty early for the convenience of the government may also receive the full thirty-six months of MGIB benefits. Depending on the reason for separation, other veterans who are separated from active duty early may be eligible for prorated (reduced) MGIB benefits—one month of benefits for each full month of active duty.

Some veterans who were eligible for the Vietnam Era GI Bill (VRA) have increased MGIB eligibility. They must have had some remaining VRA entitlement on December 31, 1989, when all benefits under the VRA expired. With some exceptions, they must have served on active duty from July 1, 1985, through June 30, 1988. For these veterans, the ten-year period of time in which they must use MGIB benefits is reduced by any time from January 1, 1977, through June 30, 1985, that they were not on active duty.

Individuals who are involuntarily separated from the military and who were not originally eligible for the MGIB may have a second opportunity to receive MGIB benefits. This includes officers not normally eligible for the MGIB because they were commissioned after December 31, 1976, as a result of graduating from a service academy or after completing a ROTC scholarship, as well as people who declined to participate in the MGIB. Contact your education center or VA for details.

$600 Buy-Up Program You can get up to $150 per month added to your standard MGIB "payment rate." This could increase your total GI Bill benefit by up to $5,400. To take advantage, you must be on active duty and elect to contribute up to $600 (in $20 increments) before you leave the service. Each $300 contributed earns an additional $75 a month in benefits. You can use DD Form 2366-1, "Increased Benefit Contribution Program," to process your request through your local payroll or personnel office (http://www.dtic.mil/whs/directives/infomgt/forms/eforms/dd2366-1.pdf).

GI Bill Apprenticeship and OJT Programs The Department of Veterans Affairs On-the-Job Training (OJT) and Apprenticeship Program offers you an alternative way to use your Montgomery GI Bill education and training benefits.

When you are trained for a new job, you can receive monthly training benefits from the VA in addition to your regular salary. This means that you can receive up to $990.75 a month (or $237.75 for Guard/Reserve) tax-free on top of your regular salary! The VA pays veteran Montgomery GI Bill participants $990.75 a month for the first six months of training, $726.55 for the second six months of training, and $462.35 for remaining training. If you are qualified for the Montgomery GI Bill or the Montgomery GI Bill for Selected Reserve and you have or are planning to start a new job or apprenticeship program, you should apply for this little-known MGIB benefit. In some cases, the VA will even pay retroactively for OJT from the past twelve months. Call 1-888-GIBILL-1 to speak to a VA representative about your eligibility for this valuable program.

Note: You may not receive GI Bill OJT benefits at the same time you receive other GI Bill education benefits.

New Post 9/11 GI Bill—Chapter 33

The Post 9/11 GI Bill is a new education benefit program that will provide service members with college tuition, stipends for housing, and books. This program went into effect on August 1, 2009. To qualify for this benefit, you must serve a minimum of ninety days on active duty after September 10, 2001. This includes active-duty service as a member of the Armed Forces or as a result of a call or order to active duty from a reserve component (National Guard and Reserve) under certain sections of Title 10. The new Post 9/11 GI Bill will pay up to 100 percent for tuition, a monthly housing stipend based on the DoD Basic Allowance for Housing at the E-5 with Dependents payment rate, and up to $1,000 a year for books and supplies.

Your benefits under the Post 9/11 GI Bill will vary depending on your state of residence, number of education units taken, and amount of post-9/11 active-duty service. Here is a quick reference showing the percentage of total combined benefit eligibility based on the following periods of post-9/11 service:

- 100 percent—thirty-six or more cumulative months
- 100 percent—thirty or more consecutive days with disability related discharge.
- 90 percent—thirty or more cumulative months
- 80 percent—twenty-four or more cumulative months
- 70 percent—eighteen or more cumulative months
- 60 percent—twelve or more cumulative months
- 50 percent—six or more cumulative months
- 40 percent—ninety or more days

However, some periods of active-duty service are excluded. Periods of service under the following do not count toward qualification for the Post 9/11 GI Bill:

- NOAA, PHS, or Active Guard Reserve
- ROTC under 10 U.S.C. 2107(b)
- Service academy contract period
- Service terminated due to defective enlistment agreement
- Service used for loan repayment
- Selected reserve service used to establish eligibility under the Montgomery GI Bill (MGIB chapter 30), MGIB for Selected Reserve (MGIB-SR chapter 1606), or the Reserve Education Assistance Program (REAP chapter 1607)

Learn more about the Post 9/11 GI Bill by downloading the Department of Veterans Affairs Post 9/11 GI Bill Pamphlet at http://www.gibill.va.gov/pamphlets/CH33/CH33_Pamphlet.pdf.

For More Information

The VA can provide you with educational counseling after you leave the Service. Contact the VA GI Bill Regional Processing Office by dialing toll-free 1-888-GI Bill-1 (1-888-442-4551) or go to the MGIB website at http://www.gibill.va.gov. To contact the VA Regional Office closest to you, go to http://www1.va.gov/directory/guide/home.asp and click on "Type of Facility." Then click on your state to locate the regional office nearest you. In addition, information on MGIB and other veterans' educational benefit programs is available from your installation's Education Center or from the admissions office and/or veterans' coordinator at most colleges and universities.

Did you know that you qualify for federal financial student aid such as Pell Grants and the Stafford Loan Program even if you are still on active duty? Visit http://www.fafsa.ed.gov/to learn how to apply.

ADDITIONAL EDUCATIONAL OR TRAINING OPTIONS

The transition from military to civilian life is an excellent time to take a serious look at your options for future success. Now is the best time to evaluate your educational options.

Guidance Counseling

Before you leave the military, go to your local Education Center, Navy College Office, or Marine Corps LifeLong Learning Center. The counselors can provide assistance in determining the goals that are right for you. If you feel you need additional education or training, the education counselor will guide you to the appropriate curriculum and institution, and help you with the paperwork necessary to enroll in an academic or vocational program.

Career Assessment

If you are not sure what you want to do upon leaving the military, you should talk to a counselor at your local Education Center, Navy College Office, Marine Corps LifeLong Learning Center, or Transition Office. The counselor can recommend aptitude tests or vocational interest inventories to help clarify your career goals. These tests can help you pinpoint job skills in which you might excel and then relate them to specific occupations and careers in the civilian world.

Your installation's Education Center, Navy College Office, or Marine Corps LifeLong Learning Center may offer the Strong Interest Inventory, Self-Directed Search, or Career Assessment Inventory, as well as computerized counseling systems like Discover. These can help you select jobs and careers that match your personality, background, and career goal.

Academic Planning

Once you have identified your career goal, you may find you need a formal education to achieve it. Your education counselor can explore the possibilities with you. Counselors can also advise you on nontraditional educational opportunities that can make it easier for you to get a diploma, vocational certificate, or college degree. These nontraditional opportunities include the following:

- Take "challenge exams," such as a college-level equivalency exam. You can convert knowledge learned outside the classroom into credits toward a college program. This can save you time and money.
- Go to school part time while continuing to hold down a full-time job. This approach might make adult education more practical.
- See the veterans' coordinator at the college, university, or vocational school of your choice. The coordinator can help you understand your VA educational benefits and might lead you to special programs offered to former service members.
- Determine if your military learning experiences can translate to course credit. Check with your service Education Center, Navy College Office, or Marine Corps LifeLong Learning Center well in advance of your separation date to request copies of your transcripts.
- Take advantage of distance learning opportunities. With today's technological advances, you can enroll in an educational program in which courses are offered by accredited educational institutions in a variety of formats, i.e., CD-ROM, the Internet, satellite TV, cable TV, and DVDs.

Vocational Services

The Education Center, Navy College Office, or Marine Corps LifeLong Learning Center can tell you about vocational and technical school programs designed to give you the skills needed to work in occupations that do not require a four-year college degree. The counselors at these centers can also show you how to get course credits for nontraditional learning experience (such as military certifications and on-the-job training). The counselors can help you explore these options. They may also help you find out about certification and licensing requirements—for example, how to get a journeyman card for a particular trade—and they can give you information on vocational and apprenticeship programs.

Note: Local trade unions may also offer vocational training in fields that interest you.

Licensing and Certification

Your military occupational specialty may require a license or certification in the civilian workforce. There are several resources available to assist you in finding civilian requirements for licensing and certification:

- Department of Labor website, http://www.acinet.org/. Go to the "Career Tools" section to look up licenses by state, requirements for the license, and point-of-contact information for the state licensing board.

- DANTES website, http://www.dantes.doded.mil/. This site has information on certification programs.
- https://www.cool.army.mil/. Find civilian credentials related to your military occupational specialty, learn what it takes to obtain the credentials, and see if there are available programs that will help pay credentialing fees.
- https://www.cool.navy.mil/. Find civilian credentials related to your Navy rating, learn what it takes to obtain the credentials, and see if there are available programs that will help pay credentialing fees.

Testing Available through Your Education Center

Testing can be an important first step in your career development. Some colleges and universities may require you to provide test results as part of your application. Prior to your departure from military service, you are encouraged to take advantage of the testing services offered by the Education Center, Navy College Office, and Marine Corps LifeLong Learning Center. These services include the following:

- *Vocational interest inventories.* Most Education Centers, Navy College Offices, and Marine Corps LifeLong Learning Centers offer free vocational interest inventories that can help you identify the careers most likely to interest you.
- *Academic entry exams.* Before applying for college or other academic programs, you may want to take a college admission test such as the Scholastic Aptitude Test (SAT), ACT, or the Graduate Record Examination (GRE). Some schools may require that you do so. Information on these tests is available from your Education Center, Navy College Office, or Marine Corps LifeLong Learning Center. You must start early. These exams are offered only a few times each year.
- *Credit by examination.* Your Education Center, Navy College Office, and Marine Corps LifeLong Learning Center offer a variety of "challenge" exams that can lead to college credit. If you score high enough, you may be exempt from taking a certain class or from certain course requirements—resulting in a big savings of time and money as you earn your degree. The College Level Examination Program (CLEP) and the DANTES Subject Standardized Tests (DSST) are also free to service members on active duty.
- *Certification examinations.* As a service member working in an important occupational field, you have received extensive training (service schools, correspondence course, OJT) that has proved valuable in developing your professional skills. Your local Education Center, Navy College Office, or Marine Corps LifeLong Learning Center can provide information on cer-

tification examinations that "translate" military training into civilian terms. Examinations are available in many skill areas, and upon successful completion the documentation you receive is readily understood and received in the professional occupational civilian community.
- *Save time and money*. You can get up to thirty college credits by taking the five CLEP General Exams. If you are currently serving in the Armed Forces, you can take these exams for free.

Contact your installation Education Center, Navy College Office, or Marine Corps LifeLong Learning Office to ensure that they have the capability to offer examinations you need in paper and pencil or computer-based-testing (CBT) format.

DoD Voluntary Education Program Website

For separating service members, the Department of Defense Voluntary Education Program website, http://www.voled.doded.mil, offers a wide variety of educational information of interest and use. The website was originally established to provide support for military education center staffs worldwide. As the website developed, it took on the mission of providing direct support to active and reserve components' service members and their families. This support includes information on all programs provided by DANTES, including the Distance Learning Program, Examination Program, Certification Program, Counselor Support Program, Troops to Teachers, and a wide variety of educational catalogs and directories.

Links are provided to each of the services' education programs and to a wide variety of education-related resources. There is also a Directory of Education Centers on the website, which contains information on all of the services' education centers worldwide, including addresses, phone numbers, and e-mail addresses.

The primary goal of the website is to provide onsite, or through links, all information for service members to select, plan, and complete their program of study, either while on active duty or upon separation.

Service Unique Transcripts

Save time and money. Unless you know for sure that you need to take a particular course, wait until the school gets *all* your transcripts before you sign up for classes. Otherwise, you may end up taking courses you don't need.

Army

For everything you want to know about the free AARTS transcript (Army/American Council on Education Registry Transcript System), go to http://aarts.army.mil. This free transcript includes your military training, your military occupational specialty (MOS), and college-level examination scores with the college credit recommended for those experiences. It is a valuable asset that you should provide to your college or your employer, and it is available for Active Army, National Guard, and Reserve Soldiers. You can view and print your own transcript at this website.

Navy and Marine Corps

Information on how to obtain the Sailor/Marine American Council on Education Registry Transcript (SMART) is available at https://www.navycollege.navy.mil/transcript.html. SMART is now able to document the American Council on Education (ACE) recommended college credit for military training and occupational experience. SMART is an academically accepted record that is validated by ACE. The primary purpose of SMART is to assist service members in obtaining college credit for their military experience. Additional information on SMART can also be obtained from your nearest Navy College Office or Marine Corps Education Center, or you can contact the Navy College Center.

Air Force

The Community College of the Air Force (CCAF) automatically captures your training, experience, and standardized test scores. Transcript information may be viewed at the CCAF website, http://www.au.af.mil/au/ccaf/.

Coast Guard

The Coast Guard Institute (CGI) requires each service member to submit documentation of all training (except correspondence course records), along with an enrollment form, to receive a transcript. Transcript information can be found at the Coast Guard Institute home page, http://www.uscg.mil/hq/cgi/.

U.S. Department of Education Financial Aid Programs

Federal Student Aid, an office of the U.S. Department of Education, offers over $80 billion in financial aid that helps millions of students manage the cost of education each year. There are three categories of federal student aid: grants, work-study, and loans. Even if you are still on active duty, you can apply for aid such as Pell Grants or federal Stafford loans. Find out more by visiting http://www.federalstudentaid.ed.gov/.

Questions and Answers about Financial Aid

Q. How do I get this aid?

By completing the Free Application for Federal Student Aid (FAFSA). You can apply online or on paper, but filing online is faster and easier. Get further instructions on the application process at http://www.fafsa.ed.gov/. You should also apply for a Federal Student Aid PIN (if you haven't done so already). The PIN allows you to sign your application electronically, which speeds up the application process even more. Apply for a PIN at http://www.pin.ed.gov/.

Q. Whose information do I include on my FAFSA?

There is a series of eight questions on the application that ask about your dependency status. If you are a veteran, or are currently serving on active duty in the U.S. Armed Forces for purposes other than training, you are considered an independent student and would only include your information (and that of your spouse, if married). For more detailed information, go to http://www.fafsa.ed.gov/.

Q. What determines my eligibility for federal student aid?

Eligibility for federal student aid is based on financial need and on several other factors. The financial aid administrator at the college or career school you plan to attend will determine your eligibility. To receive aid from federal student aid programs, you must

- demonstrate financial need (except for certain loans—your school can explain which loans are not need based)
- have a high school diploma or a General Education Development (GED) certificate, pass a test approved by the U.S. Department of Education, meet other standards your state establishes that the department approves, or complete a high school education in a home school setting that is treated as such under state law
- be working toward a degree or certificate in an eligible program
- be a U.S. citizen or eligible noncitizen
- have a valid Social Security number (unless you're from the Republic of the Marshall Islands, the Federated States of Micronesia, or the Republic of Palau)

- register with the Selective Service if required (you can use the paper or electronic FAFSA to register, you can register online at http://www.sss.gov/, or you can call 1-847-688-6888; TTY users 1-847-688-2567)
- maintain satisfactory academic progress once in school
- certify that you are not in default on a federal student loan and do not owe money on a federal student grant
- certify that you will use federal student aid only for educational purposes

Q. Can I use my Montgomery GI Bill and still get Federal Student Aid at the same time?

Yes. When you complete your FAFSA, you will be asked what you will be receiving in veterans educational benefits, which the Montgomery GI Bill falls under. Your school will take into consideration the amount you list on the application, along with any other financial assistance you are eligible to receive, in preparing your financial aid package.

Q. What if I have children who will be getting ready for college soon? Will they qualify for aid?

Federal Student Aid has a new tool called *FAFSA4caster*, designed to help students and their families plan for college. The *FAFSA4caster* provides students with an early estimate of their eligibility for federal student financial assistance. Military dependents who are enrolled in college and are eligible to receive Pell Grants should check out the two newest programs: Academic Competitiveness Grants and National Science and Mathematics Access to Retain Talent Grants (National SMART Grants). Visit their website at http://www.FederalStudentAid.ed.gov for more information.

The Veterans Upward Bound Program

The Veterans Upward Bound Program is a free U.S. Department of Education program designed to help eligible U.S. military veterans refresh their academic skills so that they can successfully complete the postsecondary school of their choosing.

The VUB program services include the following:

- basic skills development, which is designed to help veterans successfully complete a high school equivalency program and gain admission to college education programs
- short-term remedial or refresher classes for high school graduates who have put off pursuing a college education
- assistance with applications to the college or university of choice
- assistance with applying for financial aid

- personalized counseling
- academic advice and assistance
- career counseling
- assistance in getting veterans services from other available resources
- exposure to cultural events, academic programs, and other educational activities not usually available to disadvantaged people

The VUB program can help you improve your skills in:

- mathematics
- foreign language
- composition
- laboratory science
- reading
- literature
- computer basics
- any other subjects you may need for success in education beyond high school
- tutorial and study skills assistance

To be eligible for VUB, you must:

- be a U.S. military veteran with 181 or more days of active-duty service and discharged on/after January 31, 1955, under conditions other than dishonorable, *and*
- meet the criteria for low-income according to guidelines published annually by the U.S. Department of Education, *and/or* a first-generation potential college graduate, *and*
- demonstrate academic need for Veterans Upward Bound according, *and*
- meet other local eligibility criteria as noted in the local VUB project's approved grant proposal, such as county of residence, and so forth.

For more information, as well as a link to individual program locations, visit http://navub.org/.

Chapter Eight

Planning for Leaving the Military

The key to a smooth transition is to be prepared well before you separate from the military. Start early. Make connections and build networks that will help you transition smoothly into the civilian world.

OVERVIEW

There are many different sorts of transitioning—depending on whether you are single, have a family, are looking to find a civilian job, or are planning on enjoying retirement as a veteran.

The transition process is also different depending on whether you are considered a first-termer, mid-careerist, or pre-retiree; another distinction is between active duty and Reserve.

Returning to civilian life is an exciting time, one full of hope for what the next chapter in your life might bring. But the transition is also a complex undertaking. You have many steps to take, and many questions to get answered. Transition assistance staff, personnel office staff, relocation specialists, education counselors, and many others can help, but only you and your family can make the critical decisions that must be made.

If you are uncertain about your future plans, now is the time to obtain all the assistance and information you need. Professional guidance and counseling are available at your Transition Assistance Office, as are workshops, publications, information resources, automated resources, and government programs. Take advantage of each one that applies to your unique situation. It is your individual transition plan (ITP); it is your responsibility and your life.

The new "My Decision Points" ITP program will help you develop your personalized game plan for successfully transitioning back to civilian life.

"My Decision Points" provides the framework to help you identify your unique skills, knowledge, desires, experience, and abilities to help you make wise choices. It is not a Department of Defense form; it is something you create by yourself, for yourself, with information found at http://www.TurboTAP.org and assistance from a transition counselor.

TRANSITIONING ASSISTANCE

Joint Transition Assistance

The Departments of Veterans Affairs, Defense, and Labor have relaunched a new and improved website for wounded warriors: the National Resource Directory (NRD). This directory (http://www.nationalresourcedirectory.gov) provides access to thousands of services and resources at the national, state, and local levels to support recovery, rehabilitation, and community reintegration. The NRD is a comprehensive online tool available nationwide for wounded, ill, and injured service members, veterans, and their families.

The NRD includes extensive information for veterans seeking resources on VA benefits, including disability benefits, pensions for veterans and their families, VA health care insurance, and the GI Bill. The NRD's design and interface is simple, easy to navigate, and intended to answer the needs of a broad audience of users within the military, veteran, and caregiver communities.

Transition from Military to VA

VA has stationed personnel at major military hospitals to help seriously injured service members returning from Operation Enduring Freedom and Operation Iraqi Freedom (OEF/OIF) as they transition from military to civilian life. OEF/OIF service members who have questions about VA benefits or need assistance in filing a VA claim or accessing services can contact the nearest VA office or call 1-800-827-1000.

Transition Assistance Program

The Transition Assistance Program (TAP) consists of comprehensive three-day workshops at military installations designed to help service members as they transition from military to civilian life. The program includes job search, employment, and training information, as well as VA benefits information, for service members who are within twelve months of separation or twenty-four months of retirement. A companion workshop, the Disabled Transition Assistance Program, provides information on VA's Vocational Rehabilitation and Employment Program, as well as other programs for the disabled. Additional information about these programs is available at http://www.dol.gov/vets/programs/tap/tap_fs.htm.

Transition Assistance Program (TAP)

TAP consists of four essential components:

- Preseparation counseling
- DOL TAP employment workshops
- VA benefits briefings
- Disabled Transition Assistance Program (DTAP)

Pre-Discharge Program

The Pre-Discharge Program is a joint VA and DoD program that affords service members the opportunity to file claims for disability compensation and other benefits up to 180 days prior to separation or retirement.

The two primary components of the Pre-Discharge Program, Benefits Delivery at Discharge (BDD) and Quick Start, may be utilized by all separating CONUS service members on active duty, including members of the Coast Guard and members of the National Guard and Reserves (activated under Titles 10 or 32).

BDD is offered to accelerate receipt of VA disability benefits, with a goal of providing benefits within sixty days after release or discharge from active duty.

To participate in the BDD program, service members must:

1. have at least 60 days, but not more than 180 days, remaining on active duty
2. have a known date of separation or retirement
3. provide VA with service treatment records (originals or photocopies)
4. be available to complete all necessary examinations prior to leaving the point of separation

Quick Start is offered to service members who have less than sixty days remaining on active duty or are unable to complete the necessary examinations prior to leaving the point of separation.

To participate in the Quick Start Program, service members must:

1. have at least one day remaining on active duty
2. have a known date of separation or retirement
3. provide VA with service treatment records (originals or photocopies)

Service members should contact the local Transition Assistance Office or Army Career Alumni Program Center to schedule appointments to attend VA benefits briefings and learn how to initiate a pre-discharge claim. Service members can obtain more information by calling VA toll-free at 1-800-827-1000 or by visiting http://www.vba.va.gov/predischarge.

Federal Recovery Coordination Program

The Federal Recovery Coordination Program, a joint program of DoD and VA, helps coordinate and access federal, state, and local programs, benefits, and services for seriously wounded, ill, and injured service members and their families, through recovery, rehabilitation, and reintegration into the community.

Federal recovery coordinators (FRCs) have the delegated authority for oversight and coordination of the clinical and nonclinical care identified in each client's Federal Individual Recovery Plan (FIRP). Working with a variety of case managers, FRCs assist their clients in reaching their FIRP goals. FRCs remain with their clients as long as they are needed regardless of the client's location, duty status, or health status. In doing so, they often serve as the central point of contact and provide transition support for their clients.

In coordination with the Department of Defense and the Department of Health and Human Services, the joint Federal Recovery Coordinator Program is designed to cut across bureaucratic lines and reach into the private sector as necessary to identify services needed for seriously wounded and ill service members, veterans, and their families.

A key recommendation of a presidential commission chaired by former senator Bob Dole and former Health and Human Services secretary Donna Shalala, the recovery coordinators do not directly provide care, but coordinate federal health care teams and private community resources to achieve the personal and professional goals of an individualized "life map" or recovery plan developed with the service member or veteran who qualifies for the federal recovery coordinator program.

Though initially based in military facilities, their work seamlessly extends into the patient's civilian life after discharge. To ensure these severely injured persons do not get lost in the system, the coordinators actively link the veteran with public and private resources that will meet their rehabilitation needs.

Participating patients include those with seriously debilitating burns, spinal cord injuries, amputations, visual impairments, traumatic brain injuries, and posttraumatic stress disorder.

While initially focused in early stages for current military hospital inpatients, the FRCP involvement is expected to be a lifetime commitment to veterans and their families. The coordinators will maintain contacts by phone, visits, and e-mail.

When a veteran settles in a remote area, VA will use multimedia systems that integrate video and audio teleconferencing so that veterans may visit a federal clinic or private center near their homes to link up with their case coordinator for a meeting.

To get into the Federal Recovery Coordination Program, you must be seriously wounded, ill, or injured and be referred. You are referred into the program by a member of your multidisciplinary team, your commander, wounded warrior program, or through self-referral.

An assigned federal recovery coordinator will develop a FIRP with input from the service member or veteran's multidisciplinary heath care team, the service member or veteran, and his or her family or caregiver. They track the care, management, and transition of a recovering service member or veteran through recovery, rehabilitation, and reintegration.

Military Services Provide Pre-separation Counseling

Service members may receive pre-separation counseling twenty-four months prior to retirement or twelve months prior to separation from active duty. These sessions present information on education, training, employment assistance, National Guard and Reserve programs, medical benefits, and financial assistance.

Verification of Military Experience and Training

The "Verification of Military Experience and Training" (VMET) document, DD Form 2586, helps service members verify previous experience and training to potential employers, negotiate credits at schools, and obtain certificates or licenses. VMET documents are available only through Army, Navy, Air Force, and Marine Corps Transition Support Offices and are intended for service members who have at least six months of active service. Service members should obtain VMET documents from their Transition Support Office within twelve months of separation or twenty-four months of retirement.

Transition Bulletin Board

To find business opportunities, a calendar of transition seminars, job fairs, information on veterans associations, transition services, and training and education opportunities, as well as other announcements, visit the website at http://www.turbotap.org.

DoD Transportal

To find locations and phone numbers of all Transition Assistance Offices, as well as mini-courses on conducting successful job-search campaigns, writing résumés, using the Internet to find a job, and links to job search and recruiting websites, visit the DoD Transportal at http://www.veteranprograms.com/index.html.

Educational and Vocational Counseling

The Vocational Rehabilitation and Employment (VR&E) Program provides educational and vocational counseling to service members, veterans, and certain dependents (U.S.C. Title 38, Section 3697) at no charge. These counseling services are designed to help an individual choose a vocational direction, determine the course needed to achieve the chosen goal, and evaluate the career possibilities open to him or her.

Assistance may include interest and aptitude testing, occupational exploration, setting occupational goals, locating the right type of training program, and exploring educational or training facilities that can be utilized to achieve an occupational goal.

Counseling services include, but are not limited to, educational and vocational counseling and guidance; testing; analysis of and recommendations to improve job-marketing skills; identification of employment, training, and financial aid resources; and referrals to other agencies providing these services.

Eligibility

Educational and vocational counseling services are available during the period the individual is on active duty with the armed forces and is within 180 days of the estimated date of his or her discharge or release from active duty. The projected discharge must be under conditions other than dishonorable.

Service members are eligible even if they are only considering whether or not they will continue as members of the armed forces. Veterans are eligible if not more than one year has elapsed since the date they were last discharged or released from active duty. Individuals who are eligible for VA education benefits may receive educational and vocational counseling at any time during their eligibility period. This service is based on having eligibility for a VA program such as Chapter 30 (Montgomery GI Bill); Chapter 31 (Vocational Rehabilitation and Employment); Chapter 32 (Veterans Education Assistance Program—VEAP); Chapter 33 (Post-9/11 GI Bill); Chapter 35 (Dependents' Education Assistance Program) for certain spouses and dependent children; Chapter 18 (Spina Bifida Program) for certain dependent children; and Chapter 1606 and 1607 of Title 10.

Veterans and service members may apply for counseling services using VA Form 28-8832, "Application for Counseling." Veterans and service members may also write a letter expressing a desire for counseling services.

Upon receipt of either type of request for counseling from an eligible individual, an appointment for counseling will be scheduled. Counseling services are provided to eligible persons at no charge.

Veterans' Workforce Investment Program

Recently separated veterans and those with service-connected disabilities, significant barriers to employment, or who served on active duty during a period in which a campaign or expedition badge was authorized can contact the nearest state employment office for employment help through the Veterans Workforce Investment Program. The program may be conducted through state or local public agencies, community organizations, or private, nonprofit organizations.

State Employment Services

Veterans can find employment information, education and training opportunities, job counseling, job search workshops, and résumé preparation assistance at state Workforce Career or One-Stop Centers. These offices also have specialists to help disabled veterans find employment.

Unemployment Compensation

Veterans who do not begin civilian employment immediately after leaving military service may receive weekly unemployment compensation for a limited time. The amount and duration of payments are determined by individual states. Apply by contacting the nearest state employment office listed in your local telephone directory.

Veterans Preference for Federal Jobs

Since the time of the Civil War, veterans of the U.S. armed forces have been given some degree of preference in appointments to federal jobs. Veterans' preference in its present form comes from the Veterans' Preference Act of 1944, as amended, and now codified in Title 5, United States Code. By law, veterans who are disabled or who served on active duty in the U.S. armed forces during certain specified time periods or in military campaigns are entitled to preference over others when hiring from competitive lists of eligible candidates, and also in retention during a reduction in force (RIF).

To receive preference, a veteran must have been discharged or released from active duty in the U.S. armed forces under honorable conditions (honorable or general discharge). Preference is also provided for certain widows and widowers of deceased veterans who died in service; spouses of service-connected disabled veterans; and mothers of veterans who died under honorable conditions on active duty or have permanent and total service-connected disabilities. For each of these preferences, there are specific criteria that must be met in order to be eligible to receive the veterans' preference.

Recent changes in Title 5 clarify veterans' preference eligibility criteria for National Guard and Reserve members. Veterans eligible for preference include National Guard and Reserve members who served on active duty as defined by Title 38 at any time in the armed forces for a period of more than 180 consecutive days, any part of which occurred during the period beginning on September 11, 2001, and ending on the date prescribed by presidential proclamation or by law as the last date of OEF/OIF. The National Guard and Reserve service members must have been discharged or released from active duty in the armed forces under honorable conditions.

Another recent change involves veterans who earned the Global War on Terrorism Expeditionary Medal for service in OEF/OIF. Under Title 5, service on active duty in the armed forces during a war or in a campaign or expedition for which a campaign badge has been authorized also qualifies for veterans' preference. Any Armed Forces expeditionary medal or campaign badge qualifies for preference. Medal holders must have served continuously for twenty-four months or the full period called or ordered to active duty.

As of December 2005, veterans who received the Global War on Terrorism Expeditionary Medal are entitled to veterans' preference if otherwise eligible. For additional information, visit the Office of Personnel Management (OPM) website at http://www.opm.gov/veterans/html/vetguide.asp#2.

Veterans' preference does not require an agency to use any particular appointment process. Agencies can pick candidates from a number of different special hiring authorities or through a variety of different sources. For example, the agency can reinstate a former federal employee; transfer someone from another agency; reassign someone from within the agency; make a selection under merit promotion procedures or through open, competitive exams; or appoint someone noncompetitively under special authority such as a Veterans Readjustment Appointment or special authority for 30 percent or more disabled veterans. The decision on which hiring authority the agency desires to use rests solely with the agency.

When applying for federal jobs, eligible veterans should claim preference on their application or résumé. Veterans should apply for a federal job by contacting the personnel office at the agency in which they wish to work. For more information, visit https://www.usajobs.gov/veterans for job openings or help creating a federal résumé.

Veterans' Employment Opportunities Act

When an agency accepts applications from outside its own workforce, the Veterans' Employment Opportunities Act of 1998 allows preference eligible candidates or veterans to compete for these vacancies under merit promotion procedures.

Veterans who are selected are given career or career-conditional appointments. Veterans are those who have been separated under honorable conditions from the U.S. armed forces with three or more years of continuous active service. For information, visit http://www.fedshirevets.gov.

The Veterans' Recruitment Appointment allows federal agencies to appoint eligible veterans to jobs without competition. These appointments can be converted to career or career-conditional positions after two years of satisfactory work. Veterans should apply directly to the agency where they wish to work. For information, go to http://www.fedshirevets.gov/.

Small Businesses

VA's Center for Veterans Enterprise helps veterans interested in forming or expanding small businesses and helps VA contracting offices identify veteran-owned small businesses. For information, write the U.S. Department of Veterans Affairs (OOVE), 810 Vermont Avenue, N.W., Washington, DC 20420-0001, call toll-free 1-866-584-2344, or visit http://www.vetbiz.gov/.

Small Business Contracts

Like other federal agencies, VA is required to place a portion of its contracts and purchases with small and disadvantaged businesses. VA has a special office to help small and disadvantaged businesses get information on VA acquisition opportunities. For information, write the U.S. Department of Veterans Affairs (OOSB), 810 Vermont Avenue, N.W., Washington, DC 20420-0001, call toll-free 1-800-949-8387, or visit http://www.va.gov/osdbu/.

PRE-SEPARATION TIMELINE

Two Years Prior to Separation (Retirees Only)

- Schedule your pre-separation counseling appointment.
- Review the DD Form 2648, "Pre-Separation Counseling Checklist." Identify individual service providers who will provide assistance.

Eighteen Months Prior to Separation (Retirees Only)

- Attend a Transition Assistance Program workshop. If a service-connected disability makes you eligible, attend the Disabled Transition Assistance Program workshop.
- Develop your Individual Transition Plan (at home, self-directed). Seek assistance from your ACAP Center counselor, if needed.
- Make fundamental life decisions (continue working, change careers, volunteer, etc.) and determine future goals.
- Capitalize on current career stability to prepare for future career goals. Identify training, education, and/or certification requirements, and determine how to achieve goals (e.g., use tuition assistance). Start classes.
- Evaluate family requirements (college tuition, eldercare for parents, etc.).
- Determine post-retirement income requirements. Project retirement take-home pay. Identify if you need to supplement retirement take-home pay.

Twelve to Twenty-Four Months Prior to Separation (Retirees Only)

- Continue training/education needed to qualify for your objective career/pursuit.
- Investigate health and life insurance alternatives, including long-term health care coverage.
- Consider whether you will take terminal leave or cash in unused leave.
- Consider retirement locations.
- Identify medical/dental problems and arrange treatment for yourself and/or your family.
- Begin networking. Track potential network contacts you have lost or may lose contact with.
- Research your Survivor Benefit Plan (SBP) options.
- Consider spouse education and career desires.
- Update legal documents (will, powers of attorney, etc.).

Twelve Months Prior to Separation (Retirees and Separatees)

- Continue training/education needed to qualify for your objective career/pursuit.
- Schedule your pre-separation counseling appointment.
- Review the DD Form 2648, "Pre-Separation Counseling Checklist." Identify individual service providers who will provide assistance.
- Develop your Individual Transition Plan (at home, self-directed). Seek assistance from your ACAP Center counselor, if needed.

- Attend a Transition Assistance Program workshop. If a service-connected disability makes you eligible, attend the Disabled Transition Assistance Program workshop.
- Establish a financial plan to make ends meet during your transition to civilian life.
- Assess your job skills and interests. To determine how they relate to today's job market, take a vocational interest inventory. Contact your installation's Education Center and ACAP Center.
- Begin researching the job market. Develop a career plan, including a list of possible employers in your career field.
- If you need additional educational or vocational training to compete in the job market, explore your options for adult education.
- Learn about the education benefits you are eligible for under the Montgomery GI Bill (MGIB). If you enrolled in the Vietnam-era GI Bill, learn how you can convert to MGIB. Contact your local Department of Veterans Affairs (VA) representative for details.
- Visit the Education Center to take academic entrance exams, college admission test, or challenge exam. Remember, this is free to service members on active duty.
- Discuss with your family possible options about your career and where to live next.
- If you need help with your finances, explore the options.
- Review and make a copy of your personnel records.
- Start developing a résumé.
- Join a professional association in your chosen career field and become involved in it.
- *(Retirees Only)* Schedule Part I of your separation physical. Part II will be scheduled upon completion of Part I.

180 Days Prior to Separation (Retirees and Separatees)

- Continue training/education needed to qualify for your objective career/pursuit.
- Research specific job possibilities, job markets, and the economic conditions in the geographic areas where you want to live.
- Contact friends in the private sector who may help you find a job. Actively network.
- Seek assistance from your ACAP counselor after completing the first draft of your résumé.
- Attend job fairs to connect with potential employers.
- Develop an alternate plan in case your first career plan falls through.
- Review and copy your medical and dental records. Get a certified true copy of each.

- Schedule medical/dental appointments, as needed.
- Visit your ACAP Center to request your DD Form 2586, "Verification of Military Experience and Training."

150 Days Prior to Separation (Retirees and Separatees)

- Continue training/education needed to qualify for your objective career/pursuit.
- Start actively applying for jobs. Make contact with employers you will interview with.
- Start assembling a wardrobe for interviewing. Check with the ACAP Center for Dress for Success information.
- Seek help if the stress of your transition to civilian life becomes too much to handle.
- If you are separating prior to fulfilling eight years of active service, you must satisfy your obligations by becoming a member of the Reserves.
- Start posting résumés to career websites.
- Research websites for posting résumés and conducting online job search (e.g., http://www.careers.org.)
- Schedule your separation physical examination.

120 Days Prior to Separation (Retirees and Separatees)

- Complete training/education needed to qualify for your objective career/pursuit.
- If you are considering federal employment, check online at https://help.usajobs.gov/index.php/Formshttps://help.usajobs.gov/ to determine the appropriate documents to submit. Explore special federal programs and hiring opportunities for veterans.
- Consider using RESUMIX, an automated tool that allows you to use an online application to create a résumé for applying for federal jobs. You can print the résumé for your use and save it to the system to retrieve and edit for future use. For some federal jobs, you may be able to submit your résumé electronically. You may obtain more information from the USAJOBS website at http://www.usajobs.gov/.
- Continue to network aggressively.
- Visit the Relocation Assistance Program Office located at your Army Community Service Center to learn about relocation options, entitlements, and assistance.
- If you live in government housing, arrange for a preinspection and obtain termination information.

- Contact appropriate offices at your installation to discuss extended medical care (if eligible) or conversion health insurance. Learn about your options for transitional health care. If you have specific questions about veterans medical care, contact the VA, use the VA website, or make an appointment with your local VA counselor.
- Research Reserve programs to continue to receive part-time benefits, earn a future retirement, and continue to grow and train in your field. Even if you have fulfilled eight years of military service, you may want to explore the option of joining the Reserves or National Guard.
- Visit the Department of Veterans Affairs website, which contains valuable information for veterans: http://www.va.gov.
- Start a subscription to a major newspaper in the area to which you plan to move. Begin replying to want ads.
- Visit and evaluate the area to which you plan to move. Attend job interviews there. Visit a private employment agency or executive recruiter in the area.
- Send out résumés and make follow-up phone calls to check if they arrived. Submit your résumé through the DoD Job Search website at http://dod.usajobs.gov.

90 Days Prior to Separation (Retirees and Separatees)

- Continue to post résumés to websites. Conduct an automated job search for you and your spouse using the ACAP On-Line, Transition Bulletin Board, DoD Transportal, DoD Job Search, the Federal Job Opportunities Listing, http://goDefense.com, and other available employment data banks.
- Continue to expand your network.
- Consult websites that helps you locate a home, realtor, or neighborhood (database of homes for sale), such as http://www.realtor.com.
- Once you have chosen where you will live next, arrange for transportation counseling. Learn about your options for shipment and storage of household goods.
- Schedule a final dental examination.
- Determine if you are eligible for separation pay.
- If you would like to update your will or if you have legal questions or problems, obtain free legal advice.

60 Days Prior to Separation (Retirees and Separatees)

- Begin planning additional visits to the area to which you plan to move.

- Continue to send out your résumé. Include in your cover letter the date you plan to move to the area.
- Continue to network at all levels.
- Choose your transitional health care option; use military medical facilities or sign up for TRICARE, if eligible.
- For detailed information about disability compensation, benefits, and programs, call the VA at 1-800-827-1000.
- *(Retirees Only)* Complete Survivor Benefit Plan paperwork.

30 Days Prior to Separation (Retirees and Separatees)

- Continue to network.
- Review your DD Form 214, "Certificate of Release or Discharge from Active Duty."
- Several government agencies offer special loans and programs for veterans. Check with your local VA office.
- If you are unemployed, you may qualify for unemployment compensation once you are a civilian. See your local state employment office for eligibility.
- Decide whether to sign up for the optional Continued Health Care Benefit Program medical coverage.
- Complete your Veteran's Affairs Disability Application (VA Form 21-526) and turn it in to the appropriate office. Check with your local ACAP Center or VA representative.
- Consider converting your Servicemen's Group Life Insurance to Veteran's Group Life Insurance (optional).
- Review worldwide relocation information on major military and associate installations for use by military personnel and their families who are in the process of relocating: http://www.militaryinstallations.dod.mil/MOS/f?p=MI:ENTRY:0.
- Website on the military health system: http://www.tricare.osd.mil.

DEVELOPING YOUR OWN INDIVIDUAL TRANSITION PLAN (ITP)

The ITP will help you identify the actions and activities associated with your transition. Consulting with a transition assistance counselor and using the DD Form 2648-1, "Pre-separation Counseling Checklist for Reserve Component Service Members Released from Active Duty," will help you determine your options. The Transition Guide will help you work through the major headings listed on the DD Form 2648 checklist. The checklist will allow you to identify the benefits and services that will help you prepare your ITP. If you require further assistance with any of the topics covered on the Transition Counseling Checklist, please refer to the appropriate chapter of the "Transition Guide for Guard and Reserve" or online resources found at http://www.dodtap.mil. If you still need assistance, contact Military OneSource at 1-800-342-9647.

If you are uncertain about your future plans, now is the time to get all the assistance and information you need. Professional guidance and counseling is available at a Transition Assistance Office, as are workshops, publications, information resources, automated resources, and government programs. Take advantage of each one that pertains to your unique situation. It is your Individual Transition Plan—it is your responsibility and your life.

"My Decision Points"

A carefully thought out Individual Transition Plan (ITP) is your game plan for a successful transition to civilian life—it is not an official form, but something you create by yourself, for yourself. Your Transition Assistance Office will give you a head start with your DD Form 2648, "Pre-Separation Counseling Checklist," which can serve as an outline for your ITP. On this checklist, you indicate the benefits and services for which you want counseling. Your Transition Assistance Office will furnish additional information and emphasize certain points for you to consider. These selected items will help you formulate your ITP. Your Transition/ACAP or Command Career counselor (Navy) will then refer you to subject experts or other resources to get answers to your questions or additional information.

> You may be whatever you resolve to be.
>
> —Stonewall Jackson

Create Your Own Individual Transition Plan

Your ITP should identify likely actions and activities associated with your transition. You can determine what these might be through consultation with a Transition/ACAP or Command Career counselor, as well as with a VA representative or DOL representative. Remember, as stated above, to be sure to use the DD Form 2648-1. Your military service has samples of ITPs that can help you. Check with your nearest military installation Transition/ACAP or Command Career counselor (Navy) to review them. You can start developing your ITP by making decisions based on these ten not-so-simple questions:

1. What are your goals after leaving the military?
2. Where do you plan to live?
3. Do you need to continue your education or training?

4. Will the job market where you plan to relocate provide you the employment you're seeking?
5. Do you have the right skills to compete for the job(s) you're seeking?
6. Will your spouse and family goals be met at your new location?
7. Are you financially prepared to transition at this time?
8. What do you plan to do for health care?
9. How will you address the need for life insurance?
10. Which benefits are you planning to use?

In addition, the TurboTAP website gives you the opportunity to develop your own ITP online through "My Decision Points," a personalized printable ITP that you can revise at any time. Learn more at http://www.dodtap.mil.

LEAVING THE SERVICE

Next stop: civilian life! But before you go, make sure your military records are in order and double-check them for errors. It is much easier to resolve problems before you leave the service. Make a complete copy of your medical records and take it with you.

Keep Important Documents in a Safe Place

You should keep your performance ratings, service-issued licenses or certifications, DD Form 2586, "Verification of Military Experience and Training," and other service documents (such as your security clearance) in a safe and permanent file. Never give away the original copy of any of these documents.

The DD Form 214, "Certificate of Release or Discharge from Active Duty," is one of the most important documents the service will ever give you. It is your key to participation in all Department of Veterans Affairs (VA) programs as well as several state and federal programs. Keep your original in a safe, fireproof place and have certified photocopies available for reference. You can replace this record, but that takes a long time—time that you may not have. Be safe. In most states, the DD Form 214 can be registered/recorded just like a land deed or other significant document. So, immediately after you separate, register your DD Form 214 with your county recorder or town hall. If you register your documents, they can later be retrieved quickly for a nominal fee. You should check whether state or local law permits public access to the recorded document. If public access is authorized and you register the DD Form 214, others could obtain a copy for an unlawful purpose (e.g., to obtain a credit card in your name). If public access is permitted and you choose not to register your DD Form 214, you should still take steps to protect it as you would any other sensitive document (wills, marriage and birth certificates, insurance policies). You may wish to store it in a safe deposit box or at some other secure location.

In addition, your local Vet Center can certify your DD Form 214 and have a copy placed on file. Find your nearest Vet Center online at http://www1.va.gov/directory/guide/vetcenter.asp.

Documents associated with any military service should be kept in your permanent file at home. This includes those documents mentioned above.

All VA forms and correspondence also should be kept in your file, including certificates of eligibility for loans, VA file number records, and other VA papers.

Documents such as marriage licenses, birth and death certificates, and divorce and adoption papers are permanent records you will need on a recurring basis. Keep these in your permanent file as well.

You and your family members should know the location of your health records, including medical history and individual immunization records. Keep a copy in a file at home, and know where the original is kept (usually in a military medical facility or doctor's office). Don't forget to keep your family current with shots and immunizations as you transition.

Insurance policies and premium payment records should be kept in your permanent file at home.

All service members and their spouses should have a will. Once prepared by your local legal services office or through your own private attorney, it should be placed in a safe location with your other important documents.

Need to Correct Your Military Record?

Each branch of military has its own procedures for correcting the military records of its members and former members. Correction of a military record may result in eligibility for VA and other benefits—such as back pay and military retirement—that the veteran (or survivors) could not otherwise get. Generally, a request for correction must be filed within three years after the discovery of the alleged error or injustice.

If you believe there is an error in your military record, apply in writing to the appropriate Service using a DD Form 149, "Application for Correction of Military or Naval Record." The form can be submitted by the veteran, survivor, or a legal representative. Get a copy from any VA office listed in the local telephone directory or download the form from http://www.archives.gov/veterans/military-service-records/correct-service-records.html.

Replacing a Lost DD Form 214, "Certificate of Release or Discharge"

You or your next-of-kin can request a copy of your DD Form 214 online by going to the National Personnel Records Center website: http://www.archives.gov/st-louis. Or you can request the DD Form 214 by mail by sending an SF 180, "Request Pertaining to Military Records," to the National Personnel Records Center. You can download an SF 180 at http://www.archives.gov/st-louis, or request the form by fax by calling the fax-on-demand system at 1-301-837-0990 from a fax machine, using the handset. Follow the voice instructions, and request document number 2255. Or you can write a letter, including the following information in your letter:

- Your full name
- Social Security number
- Current phone number (including area code)
- Approximate dates of service
- Place of discharge
- Return address
- Reason for request

Send this request to:

National Personnel Records Center
Attention: [Your Service, e.g., Army] Records
9700 Page Avenue
St. Louis, MO 63132-5000

Or you can fax your request to 314-801-9195. For immediate assistance, you can call 314-801-0800.

Chapter Nine

Life after the Military (Transition/ Retirement)

EMPLOYMENT

The Employment Assistance Hub of the TurboTAP website can help you focus on jobs that employers need to fill today and will need to fill in the near future. Career One-Stop Center staff can help you identify the geographic areas that have opportunities in your fields of interest. Your state employment office is another good resource during this phase, offering such services as job interviewing, selection and referral to openings, job development, employment counseling, career evaluation, referral to training or other support services, and testing. Your state office can also lead you to information on related jobs nearby and introduce you to their state job banks, which have listings of jobs in your state. To look for jobs across the nation, you should check the job banks available on the TurboTAP website http://www.dodtap.mil/ for employment portal access.

And don't forget your local library's reference section. Most of them are full of helpful publications relating to job searches.

To learn about intern programs, inquire at your Transition Assistance Office, your local civilian personnel office, or the state employment office. Some government-sponsored programs, such as obtaining teaching credentials, can provide income and training in exchange for guaranteed employment. Check local and base libraries and the education office for books containing intern program information.

Temporary agencies are also a great way to become familiar with a company or industry. Explore internship possibilities with private employers; many companies have such programs but do not advertise them. Don't necessarily turn down an interesting volunteer position. Volunteering increases your professional skills and can sometimes turn into a paid position.

Although it might be tempting, you don't have to take the first job that comes along. Consider the type of work, location, salary and benefits, climate, and how the opportunity will enhance your future career growth. Even if you take the first job offer, you are not necessarily locked into it. Some experts say employers are biased against hiring the unemployed. A shrewd move might be to look for a job from a job. Take a suitable position—and then quickly move on to a better one.

TRANSITIONING FOR A WOUNDED WARRIOR INTO THE VA

All branches of the military have recovery care programs to assist wounded warriors' transition into civilian life. The program ensures that all appropriate care coordination activities, both medical and nonmedical, are completed prior to transition, so that the wounded warrior continues to receive the treatment and care required.

This process includes:

- Notification to the appropriate VA point of contact (such as a Transition Patient Advocate) when the recovering service member (RSM) begins the physical disability evaluation process, as applicable.
- Scheduling initial appointments with the Veterans Health Administration system.
- Transmittal of the RSM's military service record and health record to the VA. The transmittal includes:

    1. The RSM's authorization (or that of an individual legally recognized to make medical decisions on behalf of the RSM) for the transmittal in accordance with Public Law 104-191 (Reference (k)). The RSM may have authorized release of his or her medical records if he or she applied for benefits prior to this point in the transition. If so, a copy of that authorization shall be included with the records.
    2. The RSM's address and contact information.
    3. The RSM's DD Form 214, "Certificate of Release or Discharge from Active Duty," which shall be transmitted electronically when possible, and in compliance with Reference (d).
    4. The results of any PEB.
    5. A determination of the RSM's entitlement to transitional health care, a conversion health policy, or other health benefits through the Department of Defense, as explained in section 1145 of Title 10, United States Code (U.S.C.) (Reference (l)).

6. A copy of requests for assistance from the VA, or of applications made by the RSM for health care, compensation and vocational rehabilitation, disability, education benefits, or other benefits for which he or she may be eligible pursuant to laws administered by the Secretary of Veterans Affairs.

- Transmittal of the RSM's address and contact information to the department or agency for veterans affairs of the state in which the RSM intends to reside after retirement or separation.
- Update the CRP for the RSM's transition that shall include standardized elements of care, treatment requirements, and accountability for the plan. The CRP shall also include:

  1. Detailed instructions for the transition from the DoD disability evaluation system to the VA disability system.
  2. The recommended schedule and milestones for the RSM's transition from military service.
  3. Information and guidance designed to assist the RSM in understanding and meeting the schedule and milestones.

The RCC and RT shall:

1. Consider the desires of the RSM and the family or designated caregiver when determining the location of the RSM's care, treatment, and rehabilitation.
2. Coordinate the transfer to the VA by direct communication between appropriate medical and nonmedical staff of the losing and gaining facilities (e.g., MCCM to accepting physician).

TRANSITION FROM DOD CARE AND TREATMENT TO CIVILIAN CARE, TREATMENT, AND REHABILITATION

Prior to transition of the RSM to a civilian medical care facility, the RCC (assisted by the RT) shall ensure that all care coordination activities, both medical and nonmedical, have been completed, including:

1. Appointment scheduling with civilian medical care facility providers.
2. Transmittal of the RSM's health record to the civilian medical care facility. The transmittal shall include:

   - The RSM's authorization (or that of an individual legally recognized to make medical decisions on behalf of the RSM) for the transmittal in accordance with Reference (i).

- A determination of the RSM's entitlement to transitional health care, a conversion health policy, or other health benefits through the Department of Defense, as explained in section 1145 of Reference (l).

3. Transmittal of the RSM's address and contact information.
4. Preparation of detailed plans for the RSM's transition, to include standardized elements of care, treatment requirements, and accountability of the CRP.

The RCC and RT shall:

1. Consider the desires of the RSM and the family or designated caregiver when determining the location of the RSM's care, treatment, and rehabilitation.
2. Coordinate the transfer by direct communication between appropriate medical and nonmedical staff of the losing and gaining facilities (e.g., RCC to FRC, MCCM to accepting physician).

Upon medical retirement, the RSM receives the same benefits as other retired members of the military departments. This includes eligibility for participation in TRICARE and to apply for care through the VA.

An RSM who is enrolled in the RCP and subsequently placed on the temporary disability retired list shall continue to receive the support of an RCC, including implementation of the recovery plan, until such time as the wounded warrior program determines that the services and resources necessary to meet identified needs are in place through non-DoD programs.

TRANSITION SUPPORT

Transition from DoD Care

The RT shall provide transition support to the RSM and family or designated caregiver before, during, and after relocation from one treatment or rehabilitation facility to another or from one care provider to another. Transition preparation will occur with sufficient advance notice and information that the upcoming change in location or caregiver is anticipated by the RSM and family or designated caregiver, and will be documented in the CRP.

You and the Department of Veterans Affairs

The Department of Veterans Affairs is responsible for ensuring that you, as a disabled veteran, receive the care, support, and recognition that you have earned. The following information will help you gain access to the benefits and services you deserve.

Disabled Transition Assistance Program (DTAP)

DTAP is a briefing sponsored by the Department of Veteran Affairs, in conjunction with the Department of Defense. It may be offered following a VA benefits briefing, a Department of Labor employment workshop, or separately. Contact your local Transition/ACAP Office or Command Career Counselor to find out when a DTAP briefing is scheduled on your installation. If DTAP briefings are not available at your facility, the Transition Office or Family Center staff will refer you to other sources where similar information is available.

DTAP provides separating service members with specialized information about the Department of Veterans Affairs (VA) Vocational Rehabilitation and Employment (VR&E) Program, eligibility, and how to apply for benefits. Separating service members who believe they have a service-connected disability are strongly encouraged to request admission to the DTAP class through their unit commander. DTAP is also available online at http://www.vetsuccess.gov/. Some service members who are pending medical separation may be eligible to receive VR&E services prior to separation.

Service members being separated with a service-connected disability or being referred to a Physical Evaluation Board or placed in a "medical hold" status by their service should attend DTAP.

VA Vocational Rehabilitation Program

Vocational Rehabilitation and Employment (VR&E) is a program whose primary function is to help veterans with service-connected disabilities become suitably employed, maintain employment, or achieve independence in daily living.

The program offers a number of services to help each eligible disabled veteran reach his or her rehabilitation goal. These services include vocational and personal counseling, education and training, financial aid, job assistance, and if needed, medical and dental treatment. Services generally last up to forty-eight months, but they can be extended in certain instances.

If you need training, VA will pay your training costs, such as tuition and fees, books, supplies, equipment, and, if needed, special services. While you are in training, VA will also pay you a monthly benefit to help with living expenses, called a subsistence allowance. For details, visit http://www.vetsuccess.gov/.

Eligibility

Usually, you must first be awarded a monthly VA disability compensation payment. In some cases, you may be eligible if you aren't getting VA compensation. For example, if you are awaiting discharge from the service because of a disability, you may be eligible for vocational rehabilitation.

Eligibility is also based on your meeting the following conditions:

- you served on or after September 16, 1940, *and*
- your service-connected disabilities (SCD) are rated at least 20 percent disabling by VA, *and*
- you need vocational rehabilitation to overcome an employment handicap, *and*
- *it has been less than twelve years since VA notified you of your qualified SCD.*

Note: You may be entitled to Vocational Rehabilitation (VR) services if you are rated 10 percent disabled; however, it must be determined that you have a serious employment handicap (SEH).

Regardless of your SCD rating percentage, you may have longer than twelve years to use your VR benefit if certain conditions prevented you from participating in a VR program or it is determined that you have an SEH.

How to Apply

You can apply by filling out VA Form 28-1900, "Disabled Veterans Application for Vocational Rehabilitation," and mailing it to the VA regional office that serves your area. You can also apply online at http://vabenefits.vba.va.gov/vonapp.

VA Disability Benefits

Recent laws passed by Congress have made several changes in veterans' eligibility for VA medical care. Basically, these laws ensure that VA care will be continued for disabled veterans with service-connected disabilities.

Veterans with non-service-connected disabilities will also continue to receive VA medical care, but on a space-available basis, and a copayment may be charged. Laws are subject to change, and there are many applicable details. Contact the VA for the latest information on disability benefits.

Classifying Disabled Veterans

The VA makes an important distinction among veterans based on the nature of their disability. This distinction determines the cost and availability of VA medical services.

- *Service-connected disability.* Any veteran who was disabled by injury or disease incurred or aggravated during active military service in the line of duty will receive VA medical care on a mandatory basis. In general, this means that service will be provided as needed, at no cost to the veteran.
- *Non-service-connected disability.* Any veteran whose disability originated outside of active service will receive VA medical care on a discretionary basis. Examples of such disabilities might include disabling arthritis that you inherited from your parents, loss of the use of your legs after a fall during a ski vacation, contracting malaria, and so on. The VA generally provides medical care to those in the discretionary category on a space-available basis, so long as the veteran agrees to make a copayment.

Veterans with Service-Connected Disabilities

If your disability is service connected, your benefits fall within the mandatory category.

- *Outpatient care.* If you have a single disability or a combined disability rating of 50 percent or more, the VA will furnish outpatient care without limitation. If your disability rating is less than 50 percent, the VA will treat at no cost only those conditions that are service connected.
- *Hospital care.* The VA is required to provide hospital care at no cost. All medical services are covered while you are hospitalized. This coverage also may include transportation under certain circumstances.
- *Nursing home care.* The VA may or may not provide nursing home or domiciliary care, depending on your income and disability. For more information, call the VA.

Veterans with Non-Service-Connected Disabilities

If your disability is not service connected, the benefits you can receive are in the discretionary category.

- *Outpatient care.* With very few exceptions, outpatient care is provided to veterans with service-connected disabilities only. Contact your local VA office for details.
- *Hospital care.* Hospital care in VA facilities may or may not be provided to veterans in the discretionary category, depending on whether space and

resources are available. However, you must agree to pay a deductible of what you would pay under Medicare.
- *Nursing home care.* The VA may or may not provide nursing home care, depending on whether space and resources are available. However, you must pay a copayment. Contact the VA for details.

Department of Veterans' Affairs Compensation and Pension Programs

The Department of Veterans' Affairs offers the Veteran Disability Compensation and Veteran Pension programs, which may provide you with assistance based on your personal circumstances.

Veteran Disability Compensation

If you are a military veteran with a service-related disability, you may qualify for monthly benefit payments. These benefits are paid to veterans who are disabled by an injury or disease that occurred while on active duty (or active duty for training) or was made worse by active military service. These benefits are tax-free.

You may be eligible for disability compensation if you have a service-related disability and you were discharged under other than dishonorable conditions.

The amount of compensation that can be paid through this program ranges from $115 to $2,471 per month, depending on the severity of your disabilities.

Your monthly compensation rate is also based on other circumstances. For example, you may receive an increased payment if you have any of the following:

- very severe disabilities or loss of limb(s)
- a spouse, child(ren), or dependent parent(s)
- a seriously disabled spouse

You can apply for compensation benefits by filling out VA Form 21-526, "Veterans Application for Compensation and/or Pension" (http://www.vba.va.gov/pubs/forms/21-526.pdf). Be sure to attach copies of any of the following documents to your application:

- Discharge or separation papers (DD 214 or equivalent).
- National Guard service members should also include a copy of their military orders, presidential proclamation, or executive order that clearly demonstrates the federal nature of the service.

- Dependency records (marriage and children's birth certificates)—as applicable.
- Medical evidence (doctor and hospital reports).

You can also apply online through the VA website at http://vabenefits.vba.va.gov/vonapp.

Veteran Pension

If you are a wartime veteran with limited income *and* you are permanently and totally disabled *or* age sixty-five or older, you may be eligible for a Veteran Pension. The Veteran Pension (also known as VA Pension) is a non-service-connected benefit that provides a monthly payment to supplement your income.

You may be eligible if you meet the following criteria:

- you were discharged from service under other than dishonorable conditions, *and*
- you served ninety days or more of active duty and at least one day of that service occurred during a period of war, *and*
- your countable family income is below a yearly limit set by law, *and*
- you are permanently and totally disabled, *or*
- you are age sixty-five or older.

Note: Anyone who enlisted after September 7, 1980, generally must have served at least twenty-four months or the full period for which called or ordered to active duty. Military service from August 2, 1990, through a date to be set by law or presidential proclamation is considered to be a period of war (Gulf War). VA Pension pays you the difference between your countable family income and the yearly income limit.

This difference is generally paid in twelve equal monthly payments, rounded down to the nearest dollar.

Example: Joe (a single veteran) has an annual income of $5,000. His annual income limit is $10,929. To determine Joe's pension, his annual income of $5,000 is subtracted from the $10,929 income limit, which gives him an annual pension rate of $5,929. This translates into a monthly pension check of $494.

How to Apply You can apply for this benefit by filling out VA Form 21-526, "Veterans Application for Compensation and/or Pension" (http://www.vba.va.gov/pubs/forms/21-526.pdf). Be sure to attach copies of any of the following documents to your application:

- Discharge or separation papers (DD 214 or equivalent)
- Dependency records (marriage and children's birth certificates)—as applicable
- Medical evidence (doctor and hospital reports)

You can also apply online through the website http://vabenefits.vba.va.gov/vonapp.

Social Security Administration Benefits for Wounded Warriors

Service members can receive expedited processing of disability claims from Social Security. Benefits available through Social Security are different from those from the Department of Veterans Affairs and require a separate application.

The expedited process is used for military service members who become disabled while on active military service on or after October 1, 2001, regardless of where the disability occurs. To learn more about this benefit, visit the Social Security Wounded Warriors website at http://www.socialsecurity.gov/woundedwarriors.

Life Insurance Coverage for Service-Connected Disabled Veterans

In addition to the extended SGLI coverage and VGLI programs mentioned in Chapter 8 of the Pre-Separation Guide, veterans with service-connected disabilities are eligible for two additional life insurance programs. The following information will help you determine if you are eligible for these programs and how to apply.

Service-Disabled Veterans Life Insurance (S-DVI) S-DVI is life insurance for veterans who receive a service-connected disability rating by the Department of Veterans Affairs. The basic S-DVI program, commonly referred to as "RH Insurance," insures eligible veterans for up to $10,000 of coverage. Veterans who have the basic S-DVI coverage and are totally disabled are eligible to have their premiums waived. If waiver is granted, totally disabled veterans may apply for additional coverage of up to $20,000 under the Supplemental S-DVI program. Premiums for Supplemental S-DVI coverage, however, cannot be waived. You are eligible for S-DVI if:

- you were released from service under other than dishonorable conditions on or after April 25, 1951, *and*
- VA has notified you that you have a service-connected disability, *and*
- you are healthy except for your service-related disability, *and*
- you apply within two years of being notified of your service-connected disability.

You are eligible for SUPPLEMENTAL S-DVI if:

- you have an S-DVI policy, *and*
- the premiums on your basic coverage are being waived due to total disability, *and*
- you apply within one year of being notified of the waiver, *and*
- you are under sixty-five years of age.

The S-DVI premiums varies depending on your age, type of plan (term or permanent), and the amount of coverage you select.

You may apply for S-DVI using the following forms:

- VA Form 29-4364, "Application for Service-Disabled Veterans Insurance," to apply for basic S-DVI (http://www.insurance.va.gov/inForceGliSite/forms/29-4364.pdf)
- VA Form 29-357, "Claim for Disability Insurance Benefits," to apply for a total disability waiver of S-DVI premiums (http://www.insurance.va.gov/inforceGLISite/forms/29-357.pdf)

You can also use the Department of Veteran Affairs "Autoform" online application process, which can be found at http://www.insurance.va.gov/inforceGLISite/forms/sdviQuest/Q1a.htm.

For more information, call toll-free 1-800-669-8477 or go to http://www.insurance.va.gov.

You may be eligible for a waiver of premiums if you become totally disabled before your sixty-fifth birthday and stay that way for at least six consecutive months. (Premiums for Supplemental S-DVI can't be waived.)

Veterans' Mortgage Life Insurance (VMLI)

VMLI is an insurance program that provides up to $90,000 in mortgage life insurance coverage on the home mortgages of veterans with severe service-connected disabilities who

- receive a Specially-Adapted Housing Grant from VA for assistance in building, remodeling, or purchasing an adapted home, *and*
- have title to the home, *and*
- have a mortgage on the home.

The insurance is payable only to the mortgage lender, not to family members. VMLI coverage is available on a new mortgage, an existing mortgage, a refinanced mortgage, or a second mortgage.

VMLI premiums are determined by

- the insurance age of the veteran, *and*
- the outstanding balance of the mortgage at the time of application, *and*
- the remaining length of the mortgage.

Note: You can determine your premium rate online at https://insurance.va.gov/inForceGliSite/VMLICalc/VMLICalc.asp.

Veterans can apply by submitting VA Form 29-8636, "Veterans Mortgage Life Insurance Statement" (www.insurance.va.gov/inforceGLISite/forms/29-8636.pdf).

For more information on the VMLI program, call toll-free 1-800-669-8477 or go to http://www.insurance.va.gov.

CHAMPVA: MEDICAL CARE FOR FAMILY MEMBERS AND SURVIVORS

The Civilian Health and Medical Program of the Department of Veterans Affairs (CHAMPVA) helps pay for medical services and supplies that veterans' family members and survivors obtain from civilian sources.

To qualify, family members and survivors must *not* be eligible for Medicare or TRICARE. The following are eligible for CHAMPVA:

- the spouse or child of a veteran who has a permanent and total service-connected disability
- the surviving spouse or child of a veteran who died as a result of a service-connected condition
- the surviving spouse or child of a person who died while on active military service in the line of duty

For details and submitting new health care claims, contact:

VA Health Administration Center
CHAMPVA
P.O. Box 65024
Denver, CO 80206-9024
Toll free: 1-800-733-8387

The CHAMPVA website is www.va.gov/hac/forbeneficiaries/champva/champva.asp.

Disability Compensation

The VA pays monetary benefits to veterans who were disabled by injury or disease incurred or aggravated during active military service in the line of duty. Filing a claim with the VA (VA Form 21-256, "Application for Disability Compensation or Pension Benefits") is very important. It serves to notify the VA about your health problems, so that service-connected disabilities can be evaluated.

Note: Service members who are medically separated from the military with severance pay, and who are subsequently awarded disability compensation from the VA, will have their disability compensation offset until their severance pay has been recouped. Call the VA for details, at 1-800-827-1000.

VETERAN CENTERS

Vet Centers provide readjustment counseling and outreach services to all veterans who served in any combat zone. Services are also available for their family members for military-related issues. Veterans have earned these benefits through their service, and all are provided at no cost to the veteran or family.

Readjustment counseling is a wide range of services provided to combat veterans in the effort to make a satisfying transition from military to civilian life. Services include the following:

- Individual counseling
- Group counseling
- Marital and family counseling
- Bereavement counseling
- Medical referrals
- Assistance in applying for VA benefits
- Employment counseling
- Guidance and referral
- Alcohol/drug assessments
- Information and referral to community resources
- Military sexual trauma counseling and referral
- Outreach and community education

VA's readjustment counseling is provided at community-based Vet Centers located near veterans and their families. There is no cost for Vet Center readjustment counseling.

Find your nearest Vet Center in the online Vet Center Directory at http://www1.va.gov/directory/guide/vetcenter.asp, or check your local blue pages.

The Vet Center staff is available toll free during normal business hours at 1-800-905-4675 (Eastern) and 1-866-496-8838 (Pacific).

DISABILITYINFO.GOV—THE ONLINE DISABILITY RESOURCE

The federal government has created the http://www.disabilityinfo.gov/ website, which is designed to give people with disabilities and many others access to the information and resources they need to live full and independent lives in the workplace and in their communities. Managed by the U.S. Department of Labor's Office of Disability Employment Policy (http://www.dol.gov/odep), DisabilityInfo.gov offers a broad range of valuable information, not only for people with disabilities but also for their family members, health care professionals, service providers, and many others.

Easy to navigate, DisabilityInfo.gov is organized by subject areas that include benefits, civil rights, community life, education, employment, health, housing, technology, and transportation. By selecting a category from the tabs at the top of the home page, users are directed to valuable information covering state and local resources, news and events, grants and funding, laws and regulations, and more. Several sections of the site link to disability-related programs geared toward veterans and the military community.

With twenty-one federal agencies contributing content to this website, DisabilityInfo.gov contains extensive, frequently updated information on a host of crosscutting topics. Areas of particular interest to the military community and their families include information on the availability of assistive technologies for DoD employees and service members with disabilities; links to employment programs for transitioning wounded service members; information on benefits, compensation, and health care programs; links to relocation and employment services; special needs programs for military families; and many other Department of Defense programs serving troops and their families.

DisabilityInfo.gov also offers a free subscription service where you can sign up to receive *Disability Connection*, a quarterly newsletter, as well as other e-mail alerts covering information tailored to your individual interests. Just visit http://service.govdelivery.com/service/user.html?code=USODEP to sign up.

TRANSITIONING FOR A NONWOUNDED WARRIOR

Despite the difficult economy, there are many employment opportunities, especially for veterans. Using your training in the military can help secure a specialist job in industry, and there are many grants available to allow you to get additional training if you want to acquire additional skills. You might even choose to start your own business. If so, there are many agencies and organizations ready to offer assistance.

The Small Business Administration (SBA) is an excellent place to begin, since they offer financial assistance and loan programs specifically created for the military community.

The SBA sets loan guidelines and facilitates loans by providing guarantees to lending institution partners. The SBA, however, does not make loans to small business owners. Here's how it works: You apply for an SBA-guaranteed loan at your bank. This means you are asking SBA to provide a guarantee that you will repay the loan. The SBA sets the guidelines or requirements for the loan, and your bank, which is an SBA partner lender, makes the loan to you. Since SBA guarantees repayment, the risks normally associated with lending money are greatly reduced for the bank.

Of the numerous financial assistance options available from the SBA, two loan programs have been designed for the military community: SBA Express Loans and SBA Patriot Express Pilot Loans. The SBA streamlines the process for these programs in order to expedite loans for active-duty service members, veterans, Reservists, National Guard members, military spouses, and widowed spouses of a service member or veteran who died during service or of a service-connected disability. SBA Express Loans and SBA Patriot Express Pilot Loans may be used for most business purposes, including start-up, expansion, equipment purchases, working capital, and inventory or business-occupied real-estate purchases.

SBA Express Loans

The SBA Express Loans program relies on the lender's credit analysis, forms, and application procedures. In return for this flexibility, partner lenders accept a maximum loan guarantee of 50 percent from the SBA. The primary advantage for you, the borrower, is that you receive a response to your loan application within thirty-six hours. Additional benefits you may receive from using the SBA Express Loans program include the following:

- Easier cash flow because of extended repayment terms
- Reasonable (often lower) interest rates
- Loans may be prepaid, in whole or in part, at any time
- Lenders are prohibited from charging any fees other than those required by the SBA

SBA Patriot Express Pilot Loans

The SBA Patriot Express Pilot Loans program is designed to support and assist those in the military community who want to establish or expand their small businesses. The loan is offered by the SBA's network of participating lenders nationwide, and it features the fastest turnaround time for loan approvals by streamlining both documentation and processing. Loan amounts are up to $500,000 and qualify for SBA's maximum guarantee of up to 85 percent for loans of $150,000 or less and up to 75 percent for loans from $150,000 up to $500,000.

Patriot Express loans feature SBA's lowest interest rates for business loans, generally 2.25 percent to 4.75 percent over prime, depending upon the size and maturity of the loan. Your local SBA district office will have a listing of Patriot Express lenders in the area.

In addition, the SBA offers other types of loans and financial assistance programs. The programs are many and varied, and the qualifications for each are specific. The SBA can also provide you with education, technical assistance, information, and training.

To learn more about additional opportunities for the military community available through the SBA, please visit the website at http://www.sba.gov/vets.

Veterans Business Outreach Program and Small Business Assistance for Veterans

Veterans Business Outreach Program (VBOP)

The SBA sponsors the VBOP, which provides entrepreneurial development services to eligible veterans who own or are considering starting a small business. Services such as business training, mentoring, and counseling are available at Veterans Business Outreach Centers (VBOC).

Veterans Business Outreach Center services include the following:

- One-on-one business counseling
- General business advice
- Assessment of needs and requirements
- Identifying financing resources
- Government contracting assistance
- Franchising incentives for veteran entrepreneurs
- Training and counseling for service-disabled veterans
- Self-employment workshops
- Business plan development (includes organizational structure, a strategic plan, market analysis, and a financial plan)
- Comprehensive feasibility analysis of business plan
- Mentorship

Veterans can take advantage of these services at one of the sixteen SBA-sponsored VBOCs located throughout the United States.

Online Small Business Training

In case the VBOC is too far away, the SBA also provides free online small business training. Online courses cover relevant topics for small business, such as starting a business, financing, business management, federal government contracting, business development, international trade, franchising, accounting, and marketing.

The tools and services provided by the SBA enable veterans to start, expand, and succeed at small business. The veteran, however, must take the first step to contact the SBA for assistance. To locate the closest VBOC, visit http://archive.sba.gov/aboutsba/sbaprograms/ovbd/OVBD_VBOP.html.

Online small business training is available at http://www.sba.gov/category/navigation-structure/counseling-training.

TRANSITIONING FOR A NONWOUNDED, RETIRING WARRIOR

All those years of service will now be paid back with extra benefits and programs. As a retiree you are eligible for all the same benefits as any other veteran or disabled veteran, in addition to thc following retiree benefits.

Service members who remain on active duty or serve in the Reserves or National Guard for a sufficient period of time may retire and receive retired pay. Retirees also retain the privilege to use base facilities, such as the Commissary and gym. Those members who entered service on or after August 1, 1986, and who will qualify for an active-duty retirement, may choose between two of the current three systems. Members who become disabled while on duty may be medically retired and receive a disability retirement. See the Disability Compensation Programs section below for further details.

Military enlisted retirees can work for virtually whomever they want (except for foreign governments) and work on any project or subject matter for their new employer. However, a conflict of interest may exist if you begin to interact with certain departments or agencies of the federal government. For example, if you worked for procurement during your military career, you may be prohibited from working for a company that sells supplies to your former base.

MILITARY RETIREMENT PAY

Retirement Pay and Compensation

Service members who remain on active duty or serve in the Reserves or National Guard for twenty years or longer may retire and receive retired pay.

There are currently three retirement systems to choose from. The pay and compensation administrative officer within your command can provide more information on these system options. Also, visit DoD's online retirement pay calculators to see how each retirement system will affect payouts: http://www.defenselink.mil/militarypay/retirement/calc/index.html.

Non-Disability Retirement Pay Options

- Final Pay—The Final Pay retirement system only applies to members who first entered the service before September 8, 1980.
- High-3—The High-3 Year Average retirement system applies to members who first entered the service on or after September 8, 1980, and before August 1, 1986. High-3 also applies to members who first entered the service on or after August 1, 1986, and chose to revert to the High-3 retirement plan by not accepting the Career Status Bonus (CSB).
- CSB/REDUX—This system applies only to members who first entered the service after July 31, 1986, and chose to receive the Career Status Bonus (CSB) and the REDUX retirement plan.

Disability Retirements

Service members who become wounded, ill, or injured may be medically retired. They may receive either a permanent or temporary disability retirement. A physical examination board (PEB—medical board) determines the percent disability and recommends whether the disability is permanent or needs reexamination every eighteen months up to five years, at which time a final retirement system determination is made.

See the pay and compensation administrative officer in your command or at your military treatment facility (MTF) for additional information and referrals. Visit the DoD's online retirement pay calculators to learn more regarding your retirement pay options: http://www.defenselink.mil/militarypay/retirement/calc/index.html.

HEALTH CARE FOR RETIREES

As a retiree you have several health care benefits to choose from. These include VA-provided medical benefits, TRICARE, and other supplemental health care insurance options.

Retirees Receive Care at VA Facilities

Retirees continue to be eligible for Department of Veterans Affairs (VA) medical care on a space available basis. There are many limitations and eligibility requirements. VA medical care should *not* be relied on as your only source of health care.

TRICARE: Health Care for Retirees

Retirees and their families remain eligible to use civilian health care facilities under TRICARE. TRICARE eligibility remains in force until you are sixty-five years old. When you reach age sixty-five, TRICARE ends, and you become eligible for TRICARE for Life. For information on TRICARE, contact the beneficiary service representative or health benefits adviser at your nearest military treatment facility. You can also learn more about TRICARE at http://www.tricare.osd.mil. Go to this website to find out more about TRICARE benefits for retirees age sixty-five and older.

TRICARE Retiree Dental Plan (TRDP)

TRDP provides comprehensive dental coverage for uniformed services retirees and their family members. Under contract with the U.S. Department of Defense, the Federal Services division of Delta Dental Plan of California administers the TRDP. The TRDP is a voluntary dental benefits program with enrollee-paid premiums.

Covered services under the TRDP are offered throughout the fifty states, the District of Columbia, Puerto Rico, Guam, the U.S. Virgin Islands, American Samoa, the Commonwealth of the Northern Mariana Islands, and Canada. Visit the TRICARE Retiree Dental Plan website at http://www.trdp.org for further information.

Supplemental Health Insurance for Retirees

One short stay in the hospital could offset the cost of several years of supplemental health insurance. Even though you are covered by TRICARE, a supplemental insurance policy is a good idea for retirees. Here's why:

- TRICARE does not cover all costs.
- TRICARE has a yearly deductible to be paid.
- TRICARE has a yearly cap on non-covered expenses; the cap is extremely high, and you are responsible for the cost of non-covered items up to that amount.

If you are covered by health insurance with your new employer, you may use TRICARE as your supplemental insurance for that policy. Check with your TRICARE adviser concerning your particular circumstances.

The Supplemental Health Insurance Test

Private supplemental health insurance makes sense in a variety of situations:

- Unemployed? If you remain unemployed after retirement, you should ask yourself, "Do I have sufficient health insurance and coverage for me and my family?"
- Under-insured? After retirement, did you accept a job that does not provide full medical coverage for you or your family?
- Not insured? Do you rely on limited VA medical benefits as your only source of medical care?

If you answered *yes* to any of these questions, you should consider obtaining supplemental health insurance.

Shopping for Supplemental Health Insurance

There are many places to obtain supplemental health insurance. Several fraternal associations and many commercial insurance companies offer such plans, but you should look carefully for the one that is best for you. Insurance plans vary greatly in which medical procedures are covered and the percentage the policy will pay.

When shopping for health insurance, first consider the benefits you may have as a retiree or veteran. Then purchase supplemental insurance. The trick is to find a supplemental insurance plan that covers all your anticipated needs *without* paying for benefits that duplicate what you already have.

There are five basic types of health insurance coverage:

- *Hospital expense insurance* pays for hospital bills either in part or in full. Watch out for policies that do not pay for the first eight to ten days of a hospital stay (the average hospital stay is fewer than eight days).
- *Surgical expense insurance* covers surgeon fees. Beware: For major surgeries, all of the fees may not be covered. Read the policy carefully before you sign.

- *Medical expense insurance* covers doctors' visits in the hospital, in the doctor's office, or house calls.
- *Major medical insurance* pays practically every form of hospital and outpatient care as long as a licensed physician provides the care. Most people choose major medical because it is so comprehensive. However, the payments for this type of coverage are high.
- *Disability insurance* pays a percentage of your normal income if a disability prevents you from doing your job.

When looking at your health insurance coverage, take a moment to review your insurance on your automobile, personal property, real estate, and loan payments. Insurers sometimes offer discounts to customers who purchase several types of insurance from the same company.

Disability Compensation Programs

Military retirees who have service-connected disabilities are eligible for disability compensation. The type and amount of disability compensation eligibility is based on several factors, including the nature of the service-connected disability and the nature of your retirement. You may qualify for more than one program; however, you may not receive benefits from more than one program at a time. The following will help you determine which of these benefits you qualify for and which best suit your needs.

Veteran Disability Compensation

If you are a military retiree with a service-related disability, you may qualify for monthly benefit payments. These benefits are paid to retirees who are disabled by an injury or disease that occurred while on active duty (or active duty for training) or was made worse by active military service.

As a military retiree, you may be required to waive a portion of your military retirement to receive these tax-free benefits. This reduction in military retirement pay is commonly referred to as a VA disability offset. Certain disabled military retirees may be eligible for one of the following programs that restore some or all of the VA disability offset.

Concurrent Retirement and Disability Pay (CRDP)

CRDP is a program that provides certain military retirees a monthly restoration of some or all of their VA disability offset. Qualified military retirees are those with twenty or more years of service who have a service-connected disability of 50 percent or more. When the CRDP program is fully implemented, such members will no longer have their military retirement pay reduced by the amount of their VA disability compensation.

Unlike the Combat Related Special Compensation (CRSC), CRDP is being phased in (except for retirees who have a VA-rated, service-connected disability rating of 100 percent). The program began in 2004, and the phase-in will be complete in 2014. The phase-in is progressive. In 2007, the restoration of the offset pay was approximately 50 percent; by 2009 it was nearly 85 percent complete.

You are eligible for CRDP if you meet the following criteria:

- You must be a military retiree with twenty or more years of service, including:

    1. Chapter 61 Medical Retirees with twenty years or more.
    2. National Guard and Reserve with twenty or more good years. (After drawing a retirement check at age sixty.)
    3. Temporary Early Retirement Authority (TERA) Retirees with less than twenty years of service are also eligible.

- Have a service-related VA disability rating of 50 percent or higher.

Your CRDP payment is a restoration of retired pay. It is taxed and, if applicable, subject to collection actions for alimony, child support, community property, garnishment, and government debt, just like your retired pay.

Your monthly CRDP amount cannot exceed the lesser of your monthly gross retired pay or VA waiver amount. If you are a disability retiree whose retired pay is calculated using a percentage of disability rather than your years of service, your CRDP cannot exceed the amount your gross retired pay would currently be had it been calculated at retirement using your years of service.

Fortunately, you do not need to apply—CRDP is automatic. If you qualify, you will automatically see an increase in your monthly retirement check.

Additional information is also available at the Defense Finance and Accounting Service (DFAS) website (http://www.dfas.mil/retiredpay.html) or the Office of the Secretary of Defense for Personnel Readiness website: (http://www.defenselink.mil/prhome/mppveterans.html).

As noted above, your personal CRDP payment rate is determined by several factors. If you have questions, you can contact the DFAS by calling toll free 1-800-321-1080 to learn about your personal CRDP payment rate.

Combat Related Special Compensation (CRSC)

CRSC provides military retirees a monthly compensation that replaces their VA disability offset. Qualified military retirees are those with twenty or more years of service who have combat-related VA-rated disability. Such members are now entitled to a payment that makes up for their military retirement pay being reduced by the amount of their VA disability compensation.

CRSC includes disabilities incurred as a direct result of:

- Armed conflict
- Hazardous duty
- Conditions simulating war
- An instrumentality of war

Unlike Concurrent Retirement and Disability Pay (CRDP), CRSC has no phase-in period. Once a military retiree has been determined to be qualified, he or she will receive reduced retirement pay plus an additional sum that makes up for part or all of the offset retired pay.

Your CRSC payment is based on the percentage of your disability that your service determines to have been combat related as described above. This percentage may be less than your overall VA disability rating, and consequently the CRSC payment may be less than your offset retired pay.

Your CRSC payment is not a restoration of retired pay. It is a special entitlement payment and is not taxed.

Your monthly CRSC amount cannot exceed the lesser of your monthly gross retired pay or VA waiver amount. If you are a disability retiree whose retired pay is calculated using a percentage of disability rather than your years of service, your CRSC cannot exceed the amount your gross retired pay would currently be had it been calculated at retirement using your years of service.

Unlike CRDP, CRSC is not automatic; you will need to apply to your military service. If you qualify, your service will advise you of your status.

Retired veterans with combat-related injuries must meet all of the following criteria to apply for CRSC:

- Receiving military retired pay (this includes Chapter 61 Medical, Temporary Early Retirement [TERA] Retirees, and Temporary Disabled Retirement List [TDRL] retirees)
- Have 10 percent or greater VA rated disability due to injury
- Military retired pay is reduced by VA disability payments (VA waiver)

They must also be able to provide documentary evidence that their injury was a result of one of the following:

- Training that simulates war (e.g., exercises, field training)
- Hazardous duty (e.g., flight, diving, parachute duty)
- An instrumentality of war (e.g., combat vehicles, weapons, Agent Orange)
- Armed conflict (e.g., gunshot wounds [Purple Heart], punji stick injuries)

To learn more about the specific eligibility criteria and how to apply, visit one of the following websites:

AIR FORCE CRSC
United States Air Force
Disability Division (CRSC)
550 C Street West, Suite 6
Randolph AFB, TX 78150-4708
Phone: 1-800-616-3775
Website: http://www.afpc.randolph.af.mil/library/combat.asp

ARMY CRSC
Department of the Army
U.S. Army Physical Disabilities Agency/
Combat Related Special Compensation (CRSC)
200 Stovall Street
Alexandria, VA 22332
Phone: 1-866-281-3254
Email: crsc.info@us.army.mil
Website: http://www.CRSC.army.mil

COAST GUARD CRSC
Commander (adm-1-CRSC)
U.S. Coast Guard
Personnel Command
4200 Wilson Boulevard
Arlington, VA 22203-1804
1-800-772-8274
Website: http://www.uscg.mil/hq/cgpc/adm/adm1.htm

NAVY and MARINE CORPS CRSC
Secretary of the Navy Council of Review Boards
Attn: Combat-Related Special Compensation Branch
720 Kennon Street SE, Suite 309
Washington Navy Yard, DC 20374
Website: http://www.hq.navy.mil/ncpb/CRSCB/combatrelated.htm

Additional information is also available at the Defense Finance and Accounting Service (DFAS) website (http://www.dfas.mil/retiredpay.html) or the Office of the Secretary of Defense for Personnel Readiness website (http://www.defenselink.mil/prhome/mppveterans.html).

SURVIVOR BENEFIT PLAN PROVIDES FOR YOUR FAMILY MEMBERS

All retirees may choose to participate in the Survivor Benefit Plan or the Reserve Components Survivor Benefit Plan. The Survivor Benefit Plan is designed to provide ongoing income for your spouse and minor children should you die before them. Videos on preretirement planning and the Survivor Benefit Plan may be available for viewing at home. Check with your installation's Transition Program Office.

The Survivor Benefit Plan can be very confusing. You and your spouse will need to learn as much as possible before making your final decision. In addition, your spouse's signature is required on the form.

The Defense Accounting and Finance Service (DFAS) offers a great resource to help you learn more about the cost and benefits of the SBP. Visit http://www.dod.mil/dfas/retiredpay/survivorbenefits.html to learn more.

LEGAL ASSISTANCE

Retirees should obtain legal assistance on most personal legal matters such as wills, powers of attorney, filing federal and state income taxes, and reviewing contracts. Military legal office priority is given to active duty personnel. Retirees residing overseas may have restrictions on privileges based upon Status of Forces agreements.

NATIONAL RETIREE COUNCILS

The Military Retiree Councils provide a link between members of the military retiree community—retirees, family members, and surviving spouses living throughout the world—and the leaders of their respective military service branches.

Each National Retiree Council is comprised of a board that consists of both senior NCO and officer retirees. The members represent geographic areas within the continental United States, and at least one at-large representative.

The Retiree Councils meet annually to discuss retiree benefit issues. Upon conclusion of the meeting, they report their findings directly to the appropriate member of their respective service branch. These annual reports reflect the issues of most significance to the retiree community that year.

Note: Rather than issuing an annual report, the Air Force Retiree Council issues letters to various Air Staff offices requesting support for specific items of concern. These letters are not normally made available to the public.

Visit each of the National Retiree Council websites to view their annual reports:

- The Army Chief of Staff's Retiree Council, http://www.armyg1.army.mil/rso/RetireeCouncil.asp
- The Secretary of the Navy's Navy and Marine Corps Retiree Council, https://secnavretireecouncil.lifelines.navy.mil/CouncilInformation/tabid/254/Default.aspx
- The Air Force Retiree Council, http://www.retirees.af.mil/council/
- Coast Guard Commandant's Retiree Council, http://www.uscg.mil/hq/g-w/g-wp/g-wpm/g-wpm-2/retiree/retiree.htm

Appendix

Civil Rights for Service Members

Members of the armed forces make great sacrifices in order to protect our nation. Recognizing these sacrifices, Congress has enacted a number of laws specifically designed to protect the civil rights of service members, both while they are on active duty and after they return to civilian life. These laws ensure that service members will be able to vote while they are stationed overseas, return to their civilian employment after completing their service, and have certain civil financial protections while on active duty. Additionally, service members who have been injured during their service and return to civilian life with a new disability have civil rights guaranteed by the law.

This is a brief explanation of the civil rights of service members from a Department of Justice brochure, as well as information about how to learn more or to file a complaint if you believe your rights have been violated.

EMPLOYMENT

The Uniformed Services Employment and Reemployment Rights Act (USERRA) protects the civilian employment rights of service members and veterans. Among other things, under certain conditions, USERRA requires employers to put individuals back to work in their civilian jobs after military service. If you are eligible for reemployment, you must be restored to the position and seniority-based benefits that you would have attained or, in some cases, a comparable job, had you not been absent from work to perform military service.

USERRA also protects service members from discrimination in the workplace based on their military service or affiliation. An employer may not terminate you or deny you initial employment, retention in employment, promotion, or any other benefit of employment because of your status as a service member. An employer also may not retaliate against you or any other person for asserting, or assisting with the assertion of, USERRA rights.

USERRA also entitles you to the right to continue your existing employer-based health plan coverage for up to two years while in the military, and to reinstate your health insurance once you are reemployed.

Additionally, upon reemployment following a period of absence for uniformed service, the employer must treat you as not having had a break in service for purposes of participation, vesting, and accrual of benefits in pension plans. If you are enrolled in a contributory plan, you are allowed, but not required, to make up missed contributions to the plan or elective deferrals, and receive the employer's match, if any.

Service members who believe that they have been victims of employment discrimination based on their military service may file a complaint with the Department of Labor (DOL) or file their own lawsuit in federal or state court.

It is important that you file a complaint with DOL or consult with a private attorney as soon as possible. To file a complaint under USERRA, contact your nearest Veterans' Employment and Training Service (VETS) office, which you can locate by visiting www.dol.gov/vets/aboutvets/contacts/main.htm, or calling DOL at 1-866-4USADOL (487-2365). DOL will investigate the complaint and may attempt to voluntarily resolve the complaint. If DOL cannot resolve the complaint, upon the complainant's request, DOL will forward the complaint to the Department of Justice's Civil Rights Division.

FINANCIAL PROTECTION FOR SERVICE MEMBERS

The Servicemembers Civil Relief Act (SCRA), formerly known as the Soldiers' and Sailors' Civil Relief Act (SSCRA), is a federal law that provides a range of relief to active-duty service members. The law's purpose is to postpone or suspend certain civil obligations so that members of the armed forces can focus their full attention on their military responsibilities without adverse consequences for them or their families. It covers issues such as rental agreements, security deposits, prepaid rent, eviction, installment contracts, credit card interest rates, mortgage interest rates, mortgage foreclosure, civil judicial proceedings, automobile leases, life insurance, health insurance, and income tax payments.

The types of relief provided under the law include the following:

- Reducing the rate of interest for debts incurred before entering active duty to 6 percent
- Protecting service members against default judgments, evictions, mortgage foreclosures, and repossessions of property
- Giving service members the ability to terminate residential and automobile leases

If you think your rights under the SCRA may have been violated, you should contact your nearest Armed Forces Legal Assistance Program office to see if the SCRA applies. Dependents of service members can also contact or visit local military legal assistance offices where they reside. Office locations may be found at http://legalassistance.law.af.mil/content/locator.php.

In order to have your SCRA case reviewed by the Department of Justice, you should first seek the assistance of your military legal assistance office. If that office cannot resolve the complaint, it may choose to forward the complaint to the Department of Justice.

VOTING

The right to vote is among our most basic and cherished civil rights. The Uniformed and Overseas Citizens Absentee Voting Act (UOCAVA), enacted in 1986, protects the right of service members to vote in federal elections regardless of where they are stationed. This law requires that states and territories allow members of the United States Uniformed Services and merchant marine, their family members, and United States citizens residing outside the United States to register and vote absentee in elections for federal offices.

UOCAVA was expanded significantly in 2009, when Congress passed the Military and Overseas Voter Empowerment (MOVE) Act to provide greater protections for service members, their families, and other overseas citizens. Among its key provisions, UOCAVA, as amended by the MOVE Act, allows qualified service members and overseas citizens to:

- Register to vote and request an absentee ballot simultaneously on a single federal form.
- Request and receive voter registration and absentee ballot applications and blank absentee ballots by electronic transmission, such as by fax or e-mail.
- Have their timely-requested ballots sent to them no later than forty-five days before an election for federal office (subject to certain conditions).
- Track the receipt of their absentee ballots through a free access system.
- Use a "back-up" ballot, called the Federal Write-In Absentee Ballot (FWAB), to vote for federal offices if they have made a timely application

for, but have not received, their regular ballot from their home state or territory, subject to certain conditions. The FWAB is available at military installations and embassies worldwide, and an official online version of this ballot and instructions are available on the Defense Department's voting website.

The Federal Voting Assistance Program (FVAP) in the Department of Defense actively monitors the voter registration and absentee voting opportunities provided to members of the Armed Forces and assists service members and their families to facilitate their participation in the voting process. If you have a question or believe you have been denied any of the rights guaranteed by UOCAVA, you can contact the FVAP with the details, and FVAP can forward the relevant information to the Department of Justice for assessment. The United States Attorney General is authorized to enforce UOCAVA, and the Department of Justice has filed suits against states that fail to comply with their requirements.

The FVAP website provides detailed information about voting procedures and materials. The website can be found at www.fvap.gov. You can also contact the FVAP at:

Federal Voting Assistance Program
Department of Defense
1777 North Kent Street, Suite #14003
Arlington, VA 22209
e-mail: vote@fvap.gov
Toll-free: 800-438-VOTE (8683)

AMERICANS WITH DISABILITIES

The Americans with Disabilities Act of 1990, or the ADA, gives civil rights protections to individuals with disabilities. The ADA prohibits discrimination and guarantees that people with disabilities have the same opportunities as everyone else to participate in the mainstream of American life—to enjoy employment opportunities, to purchase goods and services, and to participate in state and local government programs and services. Modeled after the Civil Rights Act of 1964, which prohibits discrimination on the basis of race, color, religion, sex, or national origin, the ADA is an "equal opportunity" law, not a benefit program entitling you to specific services or financial assistance because of your disability.

The ADA uses different standards than the military and the Department of Veterans Affairs in determining disability status. The ADA covers people with a physical or mental impairment that substantially limits one or more major life activities, such as walking, speaking, lifting, hearing, seeing, reading, eating, sleeping, concentrating, or working. Major life activities also include the operation of major bodily functions, such as brain, immune system, respiratory, neurological, digestive, and circulatory functions.

Businesses and state and local government agencies must take reasonable steps to make it possible for people with disabilities to be their employees or customers.

For more information about protections under the ADA, visit www.ada.gov, or call the ADA information line at 800-514-0301 (voice) or 800-514-0383 (TTY)

A brochure describing the rights of service members under the ADA in greater detail can be found at http://www.ada.gov/servicemembers_adainfo.html.

Resources

GENERAL

This website has links to hundreds of other websites specific to your needs.

www.ourmilitary.mil

American Red Cross

While providing services to 1.4 million active-duty personnel and their families, the Red Cross also reaches out to more than 1.2 million members of the National Guard and the Reserves and their families who reside in nearly every community in America. Red Cross workers in hundreds of chapters and on military installations brief departing service members and their families regarding available support services and explain how the Red Cross may assist them during their deployment.

Both active-duty and community-based military can count on the Red Cross to provide emergency communications that link them with their families back home, access to financial assistance, counseling, and assistance for veterans. Red Cross Service to the Armed Forces personnel work in 756 chapters in the United States, on fifty-eight military installations around the world, and with our troops overseas.

http://www.redcross.org

Deployment Health Clinical Center

Health Clinical Center of the Department of Defense: health information for clinicians, veterans, family members, and friends.

http://www.pdhealth.mil

Iraq and Afghanistan Veterans of America (IAVA)

IAVA is the voice of the 2.8 million veterans of Iraq and Afghanistan, raising awareness in the media, on Capitol Hill and among the general public. Our mission is to connect, unite and empower post-9/11 veterans. We address critical issues facing new veterans and their families, including mental health injuries, a stretched VA system, inadequate health care for female veterans and GI Bill education benefits. We also provide valuable resources and empower veterans to connect with one another, fostering a strong and lasting community.

http://www.iava.org

Marine Corps League

The Marine Corps League was founded by Maj. Gen. Commandant John A. Lejeune in 1923 and chartered by an Act of Congress on August 4, 1937. Its membership of 51,500 is comprised of honorably discharged, active-duty and reserve Marines with ninety days of service or more, and retired Marines. Contact the Marine Corps League at 1-800-625-1775, 703-207-9588, or fax at 703-207-0047.

http://www.mcleague.org

Military OneSource

Military OneSource is a "one-stop shop" for information on all aspects of military life. From information about financial concerns, parenting, relocation, emotional well-being, work, and health to many other topics, Military OneSource can provide a wealth of information. There are many informative topics on the website specific to wounded service members and families. For example, by clicking on Personal & Family Readiness and selecting Severely Injured Service Members, you can access topics such as "Coping with Compassion Fatigue," "Finding Temporary Work during a Loved One's Extended Hospitalization," and "Reestablishing Intimacy after a Severe Injury."

In addition to the comprehensive information available online, there are representatives available at the 800 number 24/7. Calling will provide you with personalized service specific to answering your needs. You can call the same representative back for continuity of service, as each person has his or her own extension. Military OneSource is closely aligned with the Military Severely Injured Center. You can call Military OneSource as a parent, spouse, or service member. The information that you need is a phone call away: 1-800-342-9647.

http://www.militaryonesource.mil

The Military Order of the Purple Heart

The Military Order of the Purple Heart provides support and services to all veterans and their families. This website includes information on VA benefits assistance, issues affecting veterans today, and links to other key websites for veterans. Call 703-642-5360.

http://www.purpleheart.org

Noncommissioned Officers Association (NCOA)

NCOA was established in 1960 to enhance and maintain the quality of life for noncommissioned and petty officers in all branches of the armed forces, National Guard, and Reserves. The association offers its members a wide range of benefits and services designed especially for current and former enlisted service members and their families. Those benefits fall into these categories: social improvement programs to help ensure your well-being during your active military career, during your transition to civilian life, and throughout your retirement; legislative representation to serve as your legislative advocate on issues that affect you and your family, through our National Capital Office in Alexandria, Virginia; and Today's Services to help save you money through merchant program discounts. Contact NCOA at 1-800-662-2620.

http://www.ncoausa.org

Returning Veterans Resource Project NW

Provides free counseling for veterans and families in Oregon.

http://www.returningveterans.com

Soldiers to Veterans

Soldiers to Veterans is a community of disabled veterans and their families coming together to turn struggles and pain into positive action. It is a safe place to ask frank questions and speak openly of experiences. Nondisabled veterans, active-duty military, their spouses, and professionals are welcome to join the forum to mentor and empower the families struggling with life after the war.

http://www.soldierstoveterans.com

Veterans Health Information

A list of links to civilian and military health care information.

http://www.va.gov/vbs/health

Veterans of Foreign Wars (VFW)

The VFW has a rich tradition of enhancing the lives of millions through its legislative advocacy program, which speaks out on Capitol Hill in support of service members, veterans, and their families, and through community service programs and special projects. From assisting service members in procuring entitlements, to providing free phone cards to the nation's active-duty military personnel, to supporting numerous community-based projects, the VFW is committed to honoring our fallen comrades by helping the living. Contact the VFW at 202-453-5230, or fax at 202-547-3196.

http://www.vfw.org

Veterans Outreach Center (VOC)

The VOC proactively seeks out veterans in need who continue to suffer in silence—battling personal wars that can be won, with our help. VOC's collaborative approach to treatment cares for the whole person; veterans receive the breadth of services needed to regain their mental, physical, and economic health; reconnect with themselves and the community; and resume productive lives.

http://www.veteransoutreachcenter.org

Women Veterans Health Program

Provides full range of medical and mental health services for women veterans.

http://www.va.gov/wvhp

Army Reserve Websites

U. S. Army Reserves: http://www.armyreserve.army.mil
Army Reserve Family Programs; Army Reserve Family and Readiness Program: http://www.arfp.org

Army National Guard Websites

Army National Guard: http://www.1800goguard.com

Guard Family Program (one stop source of information on programs, benefits, resources on National Guard family programs): http://www.guardfamily.org
Employment Support for the Guard and Reserve (ESGR): http://www.esgr.org

COUNSELING

People House: http://www.peoplehouse.org
Veterans and Families Coming Home: http://www.veteransandfamilies.org
Lost and Found Inc.: http://www.lostandfoundinc.org
Veterans Helping Veterans Now: http://www.vhvnow.org
Pikes Peak Behavioral Health Group: http://www.ppbhg.org
Art of Redirection Counseling: http://www.artofredirection.com
Give an Hour: http://www.giveanhour.org

Special Programs for OEF/OIF Wounded

Challenge Aspen: http://www.challengeaspen.org
Challenged Athlete Foundation: http://www.challengedathletes.org
Disabled Sports USA: http://www.dsusa.org
Sun Valley Adaptive Sports: http://www.highergroundsv.org
Independence Fund: http://www.independencefund.org
Sentinels of Freedom: http://www.sentinelsoffreedom.org (scholarships)
Operation First Response: http://www.operationfirstresponse.org
Wounded Marine Careers Foundation: http://www.woundedmarinecareers.org
Semper Fi Fund (must be a Marine or been attached to a Marine unit on deployment when injuries took place): http://www.semperfifund.org
Operation Family Fund: http://www.operationfamilyfund.org
Pentagon Foundation: http://www.pentagonfoundation.org
Coalition to Salute America's Heroes: http://www.saluteheroes.org
Family and Friends for Freedom Fund: http://www.injuredmarinesfund.org
Rebuild Hope: http://www.rebuildhope.org
Hearts and Horses: http://www.heartsandhorses.org
Pikes Peak Therapeutic Riding Center: http://www.pptrc.org
Cadence Riding: http://www.cadenceriding.org
Therapeutic Riding and Education Center: http://www.trectrax.org
Outdoor Buddies: http://www.outdoorbuddies.org
Wounded Warrior Project: http://www.woundedwarriorproject.org
Wounded Heroes Fund: http://www.woundedheroesfund.net

Northrop Gumman (assisting with employment): http://operationimpact.ms.northropgrumman.com
Operation Family Fund: http://www.operationfamilyfund.org
Operation First Response: http://www.operationfirstresponse.org
Lakeshore Foundation: http://www.lakeshore.org
Project Victory: http://www.tirrfoundation.org
Military One Source Wounded Warrior Resource Center: http://www.woundedwarriorresourcecenter.com
Strikeouts for Troops: http://www.strikeoutsfortroops.org
Hope for the Warriors: http://www.hopeforthewarriors.org
Bob Woodruff Foundation: http://remind.org
Family and Friends for Freedom Fund: http://www.injuredmarinesfund.org
Fisher House: http://www.fisherhouse.org
Wounded, Ill, and Injured Compensation and Benefits Handbook: http://www.transitionassistanceprogram.com

FAMILIES

National Military Family Association (NMFA)

Serving the families of those who serve, NMFA—"The Voice for Military Families"—is dedicated to serving the families and survivors of the seven uniformed services through education, information, and advocacy. NMFA is the only national organization dedicated to identifying and resolving issues of concern to military families. Contact NMFA at 1-800-260-0218, 703-931-6632, or fax at 703-931-4600.

http://www.nfma.org

Army Families Online

The well-being liaison office assists the Army leadership with ensuring the effective delivery of well-being programs in the Army.

http://www.armyfamiliesonline.org

Army Morale Welfare and Recreation

Army recreation programs.

http://www.armymwr.com

Azalea Charities

Provides comfort and relief items for soldiers, sailors, airmen, and Marines sick, injured, or wounded from service in Iraq and Afghanistan. It purchases specific items requested by Military Medical Centers, VA Medical Centers, and Fisher House rehabilitation facilities each week. It also provides financial support to CrisisLink, a hotline for wounded soldiers and their families, and Hope for the Warriors, special projects for wounded soldiers.

http://www.azaleacharities.com/about/mission.shtml

Blue Star Mothers of America

A nonprofit organization of mothers who now have, or have had, children honorably serving in the military. Their mission is "supporting each other and our children while promoting patriotism."

http://www.bluestarmothers.org

The Military Family Network

One nation, one community, making the world a home for military families.

http://www.emilitary.org

Military Connection

Comprehensive military directory providing information on job postings, job fairs, and listings.

http://www.militaryconnection.com

Military Homefront

Website for reliable quality-of-life information designed to help troops, families, and service providers.

http://www.militaryhomefront.dod.mil

The National Remember Our Troops Campaign

The National Remember Our Troops Campaign works to recognize military service members and their families by providing an official U.S. Blue or Gold Star Service Banner. The Star Service Banner displayed in the window of a home is a tradition dating back to World War I.

http://www.nrotc.org

Strategic Outreach to Families of All Reservists (SOFAR)

SOFAR helps Reservist families reduce their stress and prepare for the possibility that their Reservist or Guard member may exhibit symptoms of trauma from serving in a combat zone. The goal of SOFAR is to provide a flexible and diverse range of psychological services that foster stabilization, aid in formulating prevention plans to avoid crises, and help families manage acute problems effectively when they occur.

http://www.sofarusa.org

FAMILY ASSISTANCE

Snowball Express: http://www.snowballexpress.org
Military One Source: http://www.militaryonesource.mil
Military news with benefit information: http://www.military.com
Freedom Calls: http://www.freedomcalls.org
Social Security: http://www.ssa.gov
Hero Salute: http://www.herosalute.com
Army Long-Term Family Case: http://www.hrc.army.mil/site/active/tagd/cmaoc/altfcm/index.htm
National Military Family Association: http://www.nmfa.org
Our Military: http://www.ourmilitary.mil/index.aspx
United Services Organization: http://www.uso.org
Cell Phones for Soldiers: http://www.cellphonesforsoldiers.com
Homes for Our Troops: http://www.homesforourtroops.org
Rebuilding Together: http://www.rebuildingtogether.org
Soldiers Angels: http://www.soldiersangels.com
Tragedy Assistance Program for Survivors: http://www.taps.org
National Resource Directory: http://www.nationalresourcedirectory.org
Armed Forces Foundation: http://www.armedforcesfoundation.org
Veterans Holidays (discounted rates): http://www.veteransholidays.com
Swords to Plowshares (employment, training, health, and legal): http://www.swords-to-plowshares.org
Grand Camps (for kids and grandparents): http://www.grandcamps.org
Army One Source: http://
Assistance with résumé, job readiness training, and so forth: http://www.chooselifeinc.org
Project Sanctuary: http://www.projectsanctuary.us
Freedom Hunters: http://www.freedomhunters.org
American Red Cross: http://www.redcross.org
Military Family Network: http://www.emilitary.org
Project Focus: http://www.focusproject.org

Military Homefront: http://www.militaryhomefront.dod.mil
Military Students on the Move: http://www.militarystudent.org
Military Spouse Career Center: http://www.military.com/spouse
National Association of Child Care Resource and Referral Agencies: http://www.naccrra.org
Military Spouse Resource Center (assistance with employment, education, scholarships): http://www.milspouse.org
Women, Infants, and Children (WIC): http://www.fns.usda.gov/wic
Our Military Kids: http://www.ourmilitarykids.org

FINANCIAL ASSISTANCE

Military Installation Finder: http://www.militaryinstallations.dod.mil/ismart/MHF-MI/
Military One Source: http://militaryonesource.mil
AnnualCreditReport.Com: http://www.annualcreditreport.com
Experian National Consumer Assistance: http://www.experian.com
EQUIFAX Credit Information Service: http://www.equifax.com
TRANSUNION: http://www.transunion.com
VA Home Loan Resources: http://www.homeloans.va.gov/veteran.htm
VA Form 26-1880, "Request for Certificate of Eligibility": http://www.vba.va.gov/pubs/forms/26-1880.pdf
Get your W-2 from myPay: https://mypay.dfas.mil/mypay.aspx
Reserve Aid: http://www.reserveaid.org
Unmet Needs: http://www.vfw.org/UnmetNeeds
Impact a Hero: http://www.impactplayer.org
Salute Heroes for Wounded Warriors: http://www.saluteheroes.org
Home Front Cares: http://www.thehomefrontcares.org
USA Cares: http://www.usacares.org
Operation Home Front: http://www.operationhomefront.net
American Soldier Foundation: http://www.soldierfoundation.org
American Military Family: http://www.AMF100.org
American Legion Temporary Financial Assistance (TFA): http://www.legion.org/veterans/family/assistance
National Association of American Veterans: http://www.naavets.org
National Veterans Foundation: http://www.nvf.org/contact/rfs/index.php
Operation Helping Healing: http://www.helpingheal.org/guidelines.html
Elks Lodge (financial assistance available): http://www.elks.org

EDUCATION

VA Education Services (GI Bill): http://www.gibill.va.gov/

VA 22-1990 Application for Education Benefits: http://www.vba.va.gov/pubs/forms/22-1990.pdf

VA Regional Office Finder: http://www1.va.gov/directory/guide/home.asp

The Defense Activity for Non-Traditional Education Support (DANTES): http://www.dantes.doded.mil/dantes_web/danteshome.asp

Department of Defense Voluntary Education Program: http://www.voled.doded.mil

Army (AARTS) Transcript: http://aarts.army.mil

Navy and Marine Corps (SMART) Transcript: http://www.navycollege.navy.mil

Air Force (CCAF) Transcript: http://www.au.af.mil/au/ccaf/

Coast Guard Institute Transcript: http://www.uscg.mil/hq/cgi/forms.html

Federal Financial Student Aid: http://www.federalstudentaid.ed.gov/

Free Application for Federal Student Aid (FAFSA)—Pell Grants or Federal Stafford Loans: http://www.fafsa.ed.gov/

Veterans' Upward Bound: http://www.veteransupwardbound.org/vetub.html

EDUCATION RESOURCES

Students Seeking Disability Related Information: http://www.abilityinfo.com

Association on Higher Education and Disability (AHEAD): http://www.ahead.org

American Council on Education: http://www.acenet.edu

Comprehensive Development Center: http://dcwi.com/~cdc/Welcome.html

National Center for Learning Disabilities: http://www.ncld.org/content/view/871/456074

Online Disability Information System (ODIS): http://www.ume.maine.edu/~cci/odis

OSERS: National Institute on Disability and Rehabilitation Research (NIDRR): http://www.ed.gov/offices/OSERS/NIDRR/index.html

Rehabilitation Counseling: http://pages.prodigy.com/rehabilitation-counseling/index.htm

UNIVERSITY RESOURCES

Centennial Colleges' Centre for Students with Disabilities (CSD): http://www.cencol.on.ca/csd

Coalition of Rehab Engineering Research Orgs: http://www.crero.org

Curry School of Education: http://curry.edchool.virginia.edu
George Washington University Rehabilitation Counselor Education Programs: http://www.gwu.edu/~chaos
Iowa State University Disabled User Services: http://www.public.iastate.edu/~dus_info/homepage.html
Johns Hopkins University Physical Medicine and Rehabilitation: http://www.med.jhu.edu/rehab
Nebraska Assistive Technology Project: http://www.nde.state.ne.us/ATP/Techhome.html
Northwestern University Rehab Engineering, Prosthetics, and Orthotics: http://www.repoc.nwu.edu
Ohio State University—Disability Services: http://www.osu.edu/units/ods
Oklahoma State University National Clearing House of Rehabilitation Training Material: http://www.nchrtm.okstate.edu/
Tarleton State University: http://www.tarleton.edu
Thomas Edison State College—Distance Learning: http://www.tesc.edu
University of California at Berkeley School of Psychology: http://www-gse.berkeley.edu/program/SP/sp.html
University of California at LA Disabilities and Computing Program (DCP): http://www.dcp.ucla.edu/
University of Delaware—SEM: http://www.ece.udel.edu/InfoAccess/
University of Georgia—Disability Services: http://www.dissvcs.uga.edu/
University of Illinois at Urbana-Champaign: http://www.uiuc.edu/
University of Kansas Medical Center, School of Allied Health: http://www.kumc.edu/SAH/
University of Minnesota, Disability Services: http://disserv.stu.umn.edu/
University of New Hampshire, Institute on Disability: http://iod.unh.edu/
University of Virginia, Special Education: http://special.edschool.virginia.edu/
University of Washington, Department of Rehab Medicine, DO-IT: http://depts.washington.edu/rehab/
Victorian University, TAFE Services: http://www.adcet.edu.au/
West Virginia University Rehabilitation Research and Training Center (WVRTC): http://www.icdi.wvu.edu/
Wright State University Rehabilitation Engineering Information and Training: http://www.engineering.wright.edu/bie/rehabengr/rehabeng.html

RELOCATION

Relocation Assistance Office Locator: http://www.militaryinstallations.dod.mil/smart/MHF-MI
"Plan My Move": http://www.militaryhomefront.dod.mil/pls.htmsdb/f?p=107:1:3267731230074301
Chamber of Commerce Locator: http://www.chamberofcommerce.com

MILITARY PERSONNEL PORTALS

Army Knowledge Online (AKO): http://www.army.mil/ako
Navy Knowledge Online (NKO): http://www.nko.mil
Air Force Portal: http://www.my.af.mil
USA Travel Source: http://www.relo.usa.com
Travel and Per Diem Information: https://secureapp2.hqda.pentagon.mil.perdiem/
The "It's Your Move" Pamphlet: http://www.usapa.army.mil/pdffiles/p55_2.pdf
"Special Needs" Resources: http://www.militaryhomefront.dod.mil/

TRANSITION

Sentinels of Freedom

Sentinels of Freedom's mission is to provide life-changing opportunities for service members who have suffered severe injuries and need the support of grateful communities to realize their dreams. Unlike any other time in history, many more severely wounded are coming home faced with the challenges of putting their lives back together. Sentinels of Freedom provides "life scholarships" to help vets become self-sufficient. Sentinels succeeds because whole communities help. Local businesses and individuals not only give money, but also time, goods and services, housing, and transportation.

http://www.sentinelsoffreedom.org

Veterans and Families Coming Home

Provides resources for vets to ease their transition from military to civilian life.

http://www.veteransandfamilies.org

GENERAL TRANSITION-RELATED WEBSITES

Summary of Veterans Benefits: http://www.vba.va.gov/bln/21/index.htm
Army Career and Alumni Program (ACAP): http://www.acap.army.mil
Civilian Assistance and Re-Employment (CARE): http://www.cpms.osd.mil/care/
Department of Veterans Affairs (DVA): http://www.va.gov
Department of Veterans Affairs Locations: http://www1.va.gov/directory/guide/home.asp?isFlash=1)
Department of Labor: http://www.dol.gov
Military Home Front: http://www.militaryhomefront.dod.mil
Military Installation Locator: http://www.militaryinstallations.dod.mil/ismart/MHF-MI/
Military OneSource: http://www.militaryonesource.mil
Operation Transition: http://www.dmdc.osd.mil/ot
DoD Transportal: http://www.dodtransportal.org/
Temporary Early Retirement Authority (TERA) Program: http://www.dmdc.osd.mil/tera
National Guard Transitional Assistance Advisors: http://www.guardfamily.org/Public/Application/ResourceFinderSearch.aspx
Air Force Airman and Family Readiness Center: http://www.militaryinstallations.dod.mil
Navy Fleet and Family Support Center: http://www.fssp.navy.mil/
Marines Career Resource Management Center (CRMC)/Transition and Employment Assistance Program Center: http://www.usmc-mccs.org/tamp/index.cfm
Coast Guard Worklife Division—Transition Assistance: http://www.uscg.mil/hq/g-w/g-wk/wkw/worklife_programs/transition_assistance.htm
Family Center, Chaplain's Office, and Related Resources Finder: http://www.nvti.cudenver.edu/resources/militarybasestap.htm
Marine for Life: http://www.mfl.usmc.mil/
Military Family Network: http://www.emilitary.org/

EMPLOYMENT

Return 2 Work: http://www.return2work.org
Military Connection: http://www.militaryconnection.com
Vet Jobs: http://www.vetjobs.com
Hire America Heroes: http://www.hireamericasheroes.org
Helmets to Hardhats: http://www.helmetstohardhats.org
Enable America: http://www.enableamerica.org

Veterans Green Jobs: http://www.veteransgreenjobs.org
Americas Heroes at Work: http://www.americasheroesatwork.gov
Veteran Job Fairs: http://www.VetsJobs.net
USA Jobs: http://www.usajobs.gov
Hire a Hero: http://www.hireAhero.com
Hire Vets First: http://www.hirevetsfirst.dol.gov
Military Officers Association of America: http://www.MOAA.org
Recruit Army: http://www.RecruitArmy.com
Recruit Navy: http://www.recruitnavy.com
Recruit Airforce: http://www.recruitairforce.com
Recruit Marines: http://www.recruitmarines.com
5 Star Recruitment Career: http://www.5starrecruitment.com
American Corporate Program—Veterans Mentoring: http://acp-usa.org
Buckley AFB NAF Human Resource Office: http://www.460fss.com/460_FSS/HTML/HRO.html
Colorado Department of Labor and Employment: http://www.coloradoworkforce.com
Colorado Springs Help Wanted: http://regionalhelpwanted.com/colorado-springs-jobs
Job Bank: http://www.jobsearch.org
Colorado State Government Job Announcements: http://www.gssa.state.co.us/announce/Job+Announcements.nsf/$about?OpenAbout

MILITARY

National Archives and Records Administration: http://www.archives.gov
DEERS/RAPIDS Locator: http://www.dmdc.osd.mil/rsl/owa/home
Military Coalition: http://www.themilitarycoalition.org
Transition Assistance Program Turbo Tap: http://www.transitionassistanceprogram.com
Military Audiology: http://www.militaryaudiology.org
To transfer Military Occupation Specialty to Civilian: http://www.OnetCenter.org

EDUCATION/SCHOLARSHIPS

Reserve Officers Association

In addition to offering scholarship and loan programs to the families of its members, the Reserve Officer Association maintains a list of military dependent scholarships and scholarships for the children of deceased service members generally. Call 1-800-809-9448.

http://www.roa.org/roal/roal_detail.asp?id=806

Military.com

Military.com is a commercial, service-related organization that maintains a website offering a scholarship search function for dependents of service members as well as state-by-state education benefits listings.

http://www.military.com

Scholarships for Military Children

Scholarships for Military Children is a scholarship program that was created by the Defense Commissary Agency. Scholarships for Military Children maintains a website that provides information on and applications for scholarships funded through the manufacturers and suppliers whose products are sold at military commissaries around the globe. 888-294-8560.

http://www.MilitaryScholar.org

Navy-Marine Corps Relief Society

The Navy-Marine Corps Relief Society maintains a website for information on and applications for educational grants offered and administered by the Navy-Marine Corps Relief Society. 703-696-4960.

http://www.nmcrs.org

Hope for the Warriors

The mission of Hope for the Warriors is to enhance quality of life for U.S. service members and their families nationwide who have been adversely affected by injuries or death in the line of duty. It has developed a number of advocacy, support, and educational programs. 910-938-1817 or 877-2HOPE4W.

info@hopeforthewarriors.org

Marine Corps Scholarship Foundation

The Marine Corps Scholarship Foundation website provides information on and applications for scholarships offered by the foundation to the sons and daughters of current or former U.S. Marines, and to the children of current or former U.S. Navy Corpsmen who have served with the United States Marine Corps. New Jersey office: 800-292-7777; Virginia office: 703-549-0060.

http://www.mcsf.org

Air Force Aid Society

The Air Force Aid Society provides need-based grants of up to $1,500 to selected sons and daughters of current, former, and deceased Air Force personnel. The Air Force Aid Society website provides information on and applications for the education grants offered by the society. Call 800-429-9475 or 703-607-3072.

http://www.afas.org/body_grant.htm

Society of Daughters of the U.S. Army (DUSA) Scholarship Awards Program

The DUSA website provides information on applications for DUSA scholarships, which are offered to daughters or granddaughters of CWOs or officers of the U.S. Army who died on active duty.

http://www.odedodea.edu/students/dusa.htm

Army Emergency Relief

In addition to providing information on and applications for scholarships provided by the Army Emergency Relief to the spouses and children of deceased Army personnel, the Army Emergency Relief also maintains a listing of general financial aid links and scholarship search engines. Call 866-878-6378 or 703-428-0000.

http://www.aerhq.org/education.asp

OTHER SOURCES OF SCHOLARSHIPS FOR MILITARY CHILDREN

Fish House Foundation: http://www.fishhouse.org
Children of Fallen Heroes: http://www.cfsrf.org
Freedom Alliance: http://www.freedomalliance.org
Scholarships for Military Children: http://www.militaryscholar.org
Troops to Teachers: http://www.dantes.doded.mil

SERVICE SPECIFIC

Navy Personnel Command: http://www.npc.navy.mil
Air Force Cross Roads: http://www.afcrossroads.com
U.S. Army Human Resource Command: http://www.hrc.army.mil/indexflash.asp

U.S. Marines: http://www.marines.mil/Pages/Default.aspx
U.S. Coastguard: http://www.uscg.mil

SERVICE SPECIFIC FINANCIAL ASSISTANCE

Air Force Aid Society: http://www.afas.org (They also have a loan called the Falcon loan, which is $500 or less for emergency needs.)
Army Emergency Relief: http://www.aerhq.org
Coast Guard Mutual Assistance (Active, Reserve, and Retired): http://www.cgmahq.org
Navy-Marine Corps Relief Society Financial Assistance: http://www.nmcrs.org

HOUSING

Home for our Troops: http://www.hfotusa.org/ (They will build you a house at no cost if accepted.)
Building Homes for Heroes: http://www.buildinghomesforheroes.com
Rebuilding Together: http://www.rebuildingtogether.org (They believe we can preserve affordable homeownership and revitalize communities by providing free home modifications and repairs, making homes safer, more accessible, and more energy efficient.)
Operation Forever Free: http://www.operationforeverfree.org

LEGAL

National Veterans Legal Services Program: http://www.nvlsp.org
American Bar Association, Pro Bono Programs: http://www.abanet.org (Information on free legal services.)

Finance and Investment Glossary

Account Balance. The sum of the dollar amounts in each TSP investment fund for an individual account. The dollar amount in each investment fund on a given day is the product of the total number of shares in that fund multiplied by the share price for that fund on that day.

Account Number. The thirteen-digit number that the TSP assigns to a participant to identify his or her TSP account. The participant must use this TSP account number (or a customized user ID) in conjunction with his or her Web password to log into the "My Account" section of the TSP website, and must use this number with his or her personal identification number (PIN) to access his or her account on the Thrift-Line.

Active Investing. A strategy of buying and selling securities based on an evaluation of the factors that affect the price of the security, such as the economy, political environment, industry trends, currency movements, and so on. The objective of an active investment strategy is to outperform the market as measured by a benchmark index such as the S&P 500.

Agency Automatic (1 percent) Contributions. Contributions equal to 1 percent of basic pay each pay period, contributed to a FERS participant's TSP account by his or her agency.

Agency Matching Contributions. Contributions made by agencies to TSP accounts of FERS employees who contribute their own money to the TSP. CSRS employees and members of the uniformed services do not receive matching contributions.

Annual Additions (Section 415(c)) Limit. An annual dollar limit, established under Internal Revenue Code (IRC) § 415(c), that limits the amount of money that can be contributed to employer-sponsored plans like the TSP. (This limit includes all employee and agency contributions.)

Annuity. Guaranteed monthly income for the life of the TSP participant (or survivor if a joint annuity) after separating from federal service. These payments are issued directly by the TSP annuity provider.

Automatic Enrollment. Applies to FERS and CSRS employees hired or rehired after July 31, 2010. As a result of the Thrift Savings Plan Enhancement Act of 2009, Public Law 111-31, that was signed into law on June 22, 2009, agencies must enroll their newly hired FERS employees in the TSP. They must also automatically enroll rehired FERS and CSRS employees who have had a break in service of more than thirty days. Automatic enrollment contributions are deducted from employees' pay at a rate of 3 percent of basic pay per pay period and deposited into their TSP accounts. Automatically enrolled participants may make a contribution election at any time to change or stop their TSP contributions.

Base Pay. This is sometimes called "basic pay." Everyone on active duty receives base pay. The amount depends on your rank and how many years you've been in the military. For example, the lowest-ranking enlisted member—someone in the pay grade of E-1—with less than two years of service, makes a base pay of $1,467 per month. A four-star general (O-10) who's been in the military for thirty years takes home $17,176 per month in base pay.

Basic Pay (Civilian). This pay is defined in 5 United States Code (USC) 8331(3).

Basic Pay (Uniformed Services). This refers to compensation payable under sections 204 and 206 of USC title 37. Section 204 pay is pay for active duty; section 206 pay (e.g., inactive duty for training (IDT) pay) is pay earned by members of the Ready Reserve (including the National Guard).

Beneficiary Participant Account. TSP account established in the name of a spouse beneficiary of a deceased TSP participant.

Bond. A debt security issued by a government entity or a corporation to an investor from whom it borrows money. The bond obligates the issuer to repay the amount borrowed (and, traditionally, interest) on a stated maturity date.

Bonus Pay (Uniformed Services). Generally, a type of special pay with its own rules for TSP contribution election purposes.

Catch-Up Contribution. Contributions that are made via payroll deductions by a participant age fifty or older, which are permitted to exceed the Internal Revenue Code (IRC) elective deferral limit.

Catch-Up Contribution Limit. An annual dollar limit, established under Internal Revenue Code (IRC) § 414(v), that limits the amount of catch-up contributions that a participant age fifty or older can make to employer-sponsored plans like the TSP. It is separate from the elective deferral limit imposed on regular employee contributions.

Child Support Allowance. By regulation, military members must provide "adequate support" to their dependents. Military members who live in unaccompanied quarters (barracks) and pay court-ordered child support receive the difference between the single and dependent rate of the military transitional rate housing allowance. This pay is called "differential pay." However, in order to receive this pay, the amount of the court-ordered child support must equal or exceed the amounts authorized. If the amount of the court-ordered child support does not equal or exceed the amounts shown on the chart, the military member does not receive this allowance. The amounts payable range from $168.60 per month for a low-ranking enlisted member to $319.80 per month for a high-ranking general.

Civil Service Retirement System (CSRS). The term "CSRS" refers to the retirement system for federal civilian employees who were hired before January 1, 1984. CSRS refers to the Civil Service Retirement System, including CSRS Offset, the Foreign Service Retirement and Disability System, and other equivalent government retirement plans.

Clothing (Uniform) Allowance. Military uniforms cost big bucks. Military members are issued (given) a complete set of uniforms during initial training. After that, it's up to the military member to replace uniform items as they become unserviceable or wear out. Enlisted members are given an annual clothing allowance to help them with this requirement. The clothing allowance is usually paid annually on a member's enlistment anniversary. Those with less than three years of service receive the basic rate (on the assumption that their uniforms are still fairly new and don't need to be replaced as much). Additionally, their first annual payment will be only half of the basic rate (on the assumption that little would have to be replaced during their first six months of service). After three years of service, enlisted members receive the standard rate each year. Officers may be reimbursed up to $400 for initial purchase of required uniform items, and up to $400 per year afterward for uniform replacement.

Combat Pay. Military members who are assigned or deployed to a designated combat zone are paid a monthly special pay, known as combat pay (or imminent danger pay). The amount paid is $225 per month for all ranks. Even if a military member only spends one second in the designated combat zone, he or she receives the entire amount of the monthly combat pay for that month.

Contribution. A deposit made to the TSP by a participant through payroll deduction or on behalf of the participant by his or her agency or service.

Contribution Allocation. A participant's choice that tells the TSP how contributions, rollovers, and loan payments that are going into his or her account should be invested among the TSP funds.

Contribution Election. A request by a participant to start contributing to the TSP, to change the amount of his or her contribution to the TSP each pay period, or to terminate contributions to the TSP.

Cost of Living Allowance. Soldiers assigned to high-cost locations in the continental United States and overseas are paid a Cost of Living Allowance. This allowance offsets the higher costs of food, transportation, clothing, and other nonhousing items. Higher costs of housing are covered separately by the Basic Allowance for Housing (BAH).

Credit Risk. The risk that a borrower will not make a scheduled payment of principal and/or interest.

Currency Risk. The risk that the value of a currency will rise or fall relative to the value of other currencies. Currency risk could affect investments in the I Fund because of fluctuations in the value of the U.S. dollar in relation to the currencies of the twenty-two countries in the EAFE index.

Customized User ID. A combination of letters, numbers, and/or symbols that you can create to use instead of your TSP account number to log into the "My Account" section of the TSP website. The user ID cannot be used on the ThriftLine as a substitute for the account number.

Designation of Beneficiary. The participant's formal indication of who should receive the money in his or her account in the event of his or her death. Participants must use Form TSP-3, "Designation of Beneficiary." (A will is not valid for the disposition of a participant's TSP account.)

Disburse. To pay out money, as from the TSP.

Drill Pay. While members on active duty (full-time duty) receive base pay, members of the National Guard and military Reserves get monthly "drill pay." The amount of monthly drill pay depends on how many drill periods a person works during the month, their military rank, and the number of years they have been in the military. Most Guard and Reserve members perform one weekend of drill per month. Each weekend counts as four drill periods. A member of the National Guard or Reserves receives one day's worth of base pay for each drill period. A Guard/Reserve member in the lowest enlisted rank (E-1), with less than two years in the military, would draw $195.68 for a weekend of drill. A full-bird colonel (O-6), with more than twenty years in the military, would make $1,538.76 for a weekend of drill. When a member of the National Guard or Reserves is performing full-time duty (such as in basic training, military job school, or deployed), they receive the same pay as active-duty members.

Elective Deferral Limit. An annual dollar limit, established under the Internal Revenue Code (IRC) § 402(g), that limits the tax-deferred contributions and Roth contributions a participant can elect to make to employer-sponsored plans like the TSP. The limit can change each year.

Eligible Employer Plan. A plan qualified under Internal Revenue Code (IRC) § 401(a), including a § 401(k) plan, profit-sharing plan, defined benefit plan, stock bonus plan, and money purchase plan; an IRC § 403(a) annuity plan; an IRC § 403(b) tax-sheltered annuity; and an eligible IRC § 457(b) plan maintained by a government employer.

Enlistment and Reenlistment Bonus. New military members who enlist with a contract to be trained in and perform a job that the military considers "critically short-manned" are entitled to an enlistment bonus. The amount of the enlistment bonus is usually included in the enlistment contract and can range from $1,000 to over $50,000. Enlistment bonuses are usually paid in a single lump sum, once the member completes initial entry training (basic training and military job training), upon arrival at the first permanent duty station.

Military members who are serving in a "shortage" job and agree to reenlist in that job (or retrain into that job) for another term may receive a reenlistment bonus. The amount of this bonus can be from $1,000 to over $90,000 for a four-year reenlistment period. Unlike initial enlistment bonuses, reenlistment bonuses are usually paid in installments: one-half at the time of reenlistment, with the remainder of the bonus paid in equal yearly installments on the anniversary of the reenlistment date. If the member reenlists while in a designated combat zone, the entire amount of the reenlistment bonus is tax-free, regardless of when actually paid.

Family Separation Allowance (FSA). Military members who are assigned or deployed to a location where their spouse and/or children are not allowed to travel at government expense are entitled to a monthly Family Separation Allowance, for each month they have been forceably separated from their dependents, after the first month. The amount of the allowance is $250 per month for all ranks. The purpose of FSA is that it costs more to maintain two separate households than it costs to maintain a single residence. This includes military basic training (after thirty days), and military job school (if dependents are not authorized).

Through September 30, 1980, FSA was payable to a member serving in pay grade E-4 (over four years of service) or above as a member with dependents. Effective October 1, 1980, FSA became payable to a member serving in any grade as a member with dependents. FSA has increased significantly since the first Gulf War:

1. Effective October 1, 1985, through January 14, 1991: $60 per month.
2. Effective January 15, 1991, through December 31, 1997: $75.
3. Effective January 1, 1998, through September 30, 2002: $100.
4. Effective October 1, 2002: $250.

Warning: If the dependents are authorized to accompany the military member at government expense to the location, but the member voluntarily elects to serve an unaccompanied tour, FSA is not payable.

Effective January 1, 1998, FSA is payable to a member married to another member regardless of whether the member has any nonactive duty dependents, when all other general conditions are met and provided members were residing together immediately before being separated by reason of military orders. Not more than one monthly allowance may be paid with respect to a married military couple for any month. Payment is made to the member whose orders resulted in the separation. If both members receive orders requiring departure on the same day, then payment goes to the senior member.

Family Subsistence Supplemental Allowance. A few low-ranking enlisted members with lots of dependents may qualify for a "Family Subsistence Supplemental Allowance," of up to $1,100 per month. A few years ago, the military was embarrassed by a news report that some low-ranking military members with lots of kids apparently qualified for food stamps. Congress reacted by changing the military pay laws to add this allowance. If a military member accepts this allowance, he or she may no longer apply for food stamps.

Federal Employees' Retirement System (FERS). The term "FERS" refers to the retirement system for federal civilian employees who were hired on or after January 1, 1984. FERS refers to the Federal Employees' Retirement System, the Foreign Service Pension System, and other equivalent government retirement plans.

Fixed Income Investments. Generally refers to bonds and similar investments (considered debt instruments) that pay a fixed amount of interest.

Food Allowance. All active-duty military members receive a monthly allowance for food, called Basic Allowance for Subsistence (BAS). Commissioned and warrant officers receive $223.84 per month, while enlisted members receive a monthly food allowance of $325.04. However, lower-ranking enlisted members who live in the barracks are generally required to consume their meals in the dining facility (chow hall), so the amount of the food allowance is immediately deducted from their paychecks. Therefore, they get free meals, as long as they eat those meals in the chow hall.

Officers, enlisted members who live off-base or in family housing, and higher-ranking enlisted members do not receive free meals in the chow hall—instead they receive the monthly food allowance. If they choose to eat in the chow hall, they must pay for each meal. Those on a "meal card" (free meals in the chow hall) can claim a "missed meal" if they are not able to eat a meal in the chow hall due to duty reasons. If the commander approves the "missed meal," then the member receives the cost of that meal in his or her next paycheck.

Full Withdrawal. A post-separation withdrawal of a participant's entire TSP account through an annuity, a single payment, or TSP monthly payments (or a combination of these three options).

Housing Allowance. Military recruiters promise "free room and board." The "room" part of this promise is accomplished through the military's housing program. Enlisted members who are fairly new to the military and do not have a spouse and/or children generally live in a military barracks (dormitory). Because military barracks generally do not meet minimum military housing standards required by law, most people who live in the barracks also receive a few bucks each month for their inconvenience, in the form of Partial Housing Allowance. With the exception of basic training and military job school, the new "standards" for most of the services now include a single room for each person, with a bathroom shared with one or more others. As enlisted members progress in rank to above E-4, they are usually given the opportunity to move off-base and rent a house or apartment—receiving a monthly housing allowance. At many locations, lower-ranking enlisted members can also choose to move off-base if they wish, but it will be at their own expense.

Individuals who are married and/or reside with dependents either receive an on-base family house rent-free, or they receive a monthly housing allowance to rent (or buy) a place off-base. The amount of the monthly housing allowance depends on the member's rank, location of assignment, and whether or not he or she has dependents (spouse and/or children).

Members of the National Guard and Reserves are also entitled to a housing allowance when on full-time active duty. However, it works a little differently. If the Guard/Reserve member is on active duty (full-time duty) for thirty days or longer, they receive the same monthly housing allowance as active-duty members. However, if they perform active duty for less than thirty days, they receive a different housing allowance, which usually pays less and doesn't depend on the member's location. Guard and Reserve members do not receive a housing allowance when performing weekend drill duty.

Incentive Pay (Uniformed Services). Pay set forth in Chapter 5 of USC title 37 (e.g., flight pay, hazardous duty pay).

Index. A broad collection of stocks or bonds that is designed to match the performance of a particular market. For example, the Standard & Poor's 500 (S&P 500) is an index of large and medium-sized U.S. companies.

Index Fund. An investment fund that attempts to track the investment performance of an index.

Inflation Risk. The risk that investments will not grow enough to offset the effects of inflation.

In-Service Withdrawal. A disbursement made from a participant's account that is available only to a participant who is still employed by the federal government, including the uniformed services.

Interfund Transfer (IFT). An IFT allows the participant to redistribute all or part of his or her money already in the TSP among the different TSP funds. For each calendar month, the participant's *first two* IFTs can redistribute money in his or her account among any or all of the TSP funds. After that, for the remainder of the month, the participant's IFTs can *only* move money into the Government Securities Investment (G) Fund (in which case, the participant will increase the percentage of his or her account held in the G Fund by reducing the percentage held in one or more of the other TSP funds). An IFT does not change the way new contributions, transfers, or rollovers into the TSP, or loan payments are invested.

Investment Allocation. A participant's choice that tells the TSP (1) how money going into his or her account should be invested in the TSP funds (contribution allocation) and/or (2) how money already in the TSP account should be invested in the TSP funds (interfund transfer). An investment allocation can be made on the TSP website in "My Account," or by calling the toll-free ThriftLine at 1-TSP-YOU-FRST (1-877-968-3778). (See "Contribution Allocation" and "Interfund Transfer.")

IRS Life Expectancy Tables. When you withdraw your account, if you choose to have the TSP calculate monthly payments based on life expectancy, the TSP will use these tables. IRS Single Life Table, Treas. Reg. § 1.401(a)(9)-9, Q&A 1, is used for participants who are under age seventy on or after July 1 of the calendar year in which the calculation is made. For participants who turn age seventy before July 1 of that year, the Uniform Lifetime Table, Treas. Reg. § 1.401(a)(9)-9, Q&A 2, is used.

Job-Related Pay. Some military members receive extra pay, due to the nature of their military job or assignment:

Flight Pay. Military members who perform regular flight duties are entitled to special monthly flight pays.

Hazardous Duty Pay. Military members who perform duties that are considered "dangerous," due to the nature of the job, are entitled to monthly hazardous duty incentive pay.

Diving Duty Pay. Each branch of service has specific qualification requirements for members to receive diving duty pay. In general, these divers must be designated divers by order, training, and assignment.

Sea Pay. Military members performing duty at sea are entitled to a special monthly pay, known as "career sea pay."

Submarine Duty Pay. Military personnel (mostly Navy) who perform operational submarine duty are entitled to receive submarine duty pay.

Market Risk. The risk of a decline in the market value of stocks or bonds.

Matching Contributions. See "Agency Matching Contributions."

Mixed Withdrawal. A post-employment withdrawal of a participant's entire account through any combination of the following: an annuity, a single payment, or TSP monthly payments.

Monthly Payments. See "TSP Monthly Payments."

Moving and Relocation Allowance. The military pays for the transportation of household goods during a permanent change-of-station move. In addition, a Temporary Lodging Allowance covers the cost of temporary housing at the beginning and end of a move. To further offset moving costs, a Dislocation Allowance is also granted for permanent change-of-station moves.

My Account. The secure section of the TSP website, where you can log into your account to find out your account balance or perform certain transactions.

Nonpay Status. Actively employed by the federal government or uniformed services but not receiving regular pay because of furlough, suspension, leave without pay (including leave without pay to perform military service), or pending resolution of a grievance or appeal.

Partial Withdrawal. A one-time post-employment distribution of part of a participant's account balance. A partial withdrawal is participant-elected and is made in a single payment.

Participant Statements. Statements that are furnished to each TSP participant after the end of each calendar quarter and after the end of each calendar year. Quarterly statements show the participant's account balance (in both dollars and shares) and the transactions in his or her account during the quarter covered. Annual statements summarize the financial activity in the participant's account during the year covered and provide other important account data such as the participant's personal investment performance, primary beneficiary information, and an account profile.

Passive Investing. Generally, buying and holding a portfolio of securities designed to replicate a broad market index. Passive strategies are based on the assumption that it is impossible to accurately forecast future trends in securities prices over long periods of time. Management fees and trading costs are generally lower in passively managed index funds.

Password. A secret eight-character code made up of letters and numbers that a TSP participant uses in conjunction with his or her TSP account number (or customized user ID) whenever accessing his or her account through the TSP website. For new participants, the initial password is computer-generated and is sent to the participant shortly after his or her first contribution is received by the TSP. Participants can customize their passwords using the TSP website.

Pay Status. Actively employed by the federal government or uniformed services and receiving regular pay.

Personal Identification Number (PIN). A four-digit number that the participant can use (in conjunction with his or her TSP account number) to access his or her own account on the ThriftLine. The initial PIN is computer-generated and is sent to the participant shortly after the participant's first contribution is received by the TSP.

Post-Separation Withdrawal. A distribution from a participant's account that is available only to participants who have left federal service or the uniformed services. Sometimes referred to as a "post-employment" withdrawal. (See also "Withdrawal.")

Prepayment Risk. The probability that as interest rates fall, bonds that are represented in the index will be paid back early, thus forcing lenders to reinvest at lower rates.

Qualified Earnings. Earnings on Roth contributions that are eligible to be paid out tax-free at withdrawal. Earnings are considered "qualified" as long as the following two requirements are met: (1) it has been five years since January 1 of the calendar year the participant made the first Roth TSP contribution *and* (2) the participant is at least age 59½ and permanently disabled (or deceased).

Reamortize. Adjust the terms of a loan to change the loan payment amount or to shorten or lengthen the repayment period.

Required Minimum Distribution. The amount of money, based on a participant's age and previous year's TSP account balance, that the

IRS requires be distributed to a participant each year after the participant has reached age 70½ and is separated from service.

Risk (Volatility). The amount of change (both up and down) in an investment's value over time.

Roth Balance. The portion of your TSP account made up of Roth (after-tax) contributions and accrued earnings. Portions of this balance may have originated from tax-exempt pay.

Roth Contributions. Contributions from pay that has already been taxed (or from tax-exempt pay) and that has been deposited to a Roth balance.

Roth IRA. An individual retirement account that is described in § 408A of the Internal Revenue Code (IRC). A Roth IRA provides tax-free earnings. You must pay taxes on the funds you transfer from your traditional balance to a Roth IRA; the tax liability is incurred for the year of the transfer.

Securities. A general term describing a variety of financial instruments, including stocks and bonds.

SIMPLE IRA. Savings Incentive Match Plan for Employer, an employer-sponsored retirement plan available to small businesses. A TSP participant can transfer money from a SIMPLE IRA to the TSP, as long as he or she participated in the SIMPLE IRA for at least two years. However, a participant cannot transfer an amount from a TSP account into a SIMPLE IRA.

Single Payment. A payment made at one time. Sometimes referred to as a "lump sum."

Special Pay (Uniformed Services). Pay set forth in Chapter 5 of United States Code (USC) Title 37 (e.g., medical and dental officer pay, hardship duty pay, career sea pay).

Stocks. Equity securities issued as ownership in a publicly held corporation.

Tax Advantage. Not all military pay is subject to federal or state income tax. Because this pay goes into the military member's pocket, instead of the government's pocket, this is like getting a few bucks extra each month. In most (but not all) cases, if it's called "pay" (such as "basic pay"), it's subject to income tax. If it's called an "allowance" (such as "Basic Allowance for Housing" or "Subsistence Allowance"), it's not.

For duty performed in a designated combat zone, all income earned by enlisted members or warrant officers is tax exempt. For officers, the amount of income that is tax-exempt in a combat zone is equal to the maximum amount of base pay paid to the highest-ranking enlisted member. For 2011, that's $6,527 per month.

Tax-Exempt Contributions. Contributions that can be made to the TSP by members of the uniformed services from pay that is covered by the combat zone tax exclusion.

ThriftLine. The TSP's automated voice response system. It provides general news about the TSP and allows participants to access certain information and perform some transactions over the telephone. You can also use the ThriftLine to contact Participant Service Representatives at the TSP. To access your account through the ThriftLine, you will need your TSP account number and ThriftLine PIN.

Time Horizon. The investment time you have until you need to use your money.

Traditional Balance. The portion of your TSP account made up of your pre-tax (and any tax-exempt) TSP contributions, plus agency contributions and accrued earnings.

Traditional Contributions. Contributions from pay that has not yet been taxed. Also referred to as "tax-deferred," "pre-tax," or "non-Roth" contributions. Traditional contributions also include contributions to a traditional balance from tax-exempt pay earned in a combat zone.

Traditional IRA. A traditional individual retirement account described in § 408(a) of the Internal Revenue Code (IRC) or an individual retirement annuity described in IRC § 408(b). It does not include a Roth IRA, a SIMPLE IRA, or a Coverdell Education Savings Account (formerly known as an education IRA).

TSP Monthly Payments. Payments that the participant elects to receive each month from his or her TSP account after separating from service. (*Note*: In this case, money remains in the TSP account and is paid out directly from the account.)

Uniformed Services. Uniformed members of the Army, Navy, Air Force, Marine Corps, Coast Guard, Public Health Service, and the National Oceanic and Atmospheric Administration serving on active duty and members of the Ready Reserve or National Guard of those services in any pay status.

User ID. See "Customized User ID."

Vesting. For a FERS participant, the time in service that he or she must have upon separation from service in order to be entitled to keep Agency Automatic (1 percent) Contributions and associated earnings. A participant is vested in (entitled to keep) the Agency Automatic (1 percent) Contributions in his or her account after completing three years of federal service (two years for most FERS employees in congressional and certain noncareer positions).

Volatility. See "Risk."

Withdrawal. A general term for a distribution that a participant requests from his or her account. (Includes in-service withdrawal, partial withdrawal, full withdrawal, etc.)

Index

About the Authors

Cheryl Lawhorne-Scott is a clinical therapist with an eighteen-year track record of counseling services specializing in trauma care, posttraumatic stress, and traumatic brain injury treatment for wounded, ill, and injured service members and their families. As a senior consultant under the Office of the Secretary of Defense, she is part of a team that seeks innovative and proactive ways to enhance resources and services to military members and their families. She recently participated in the corporate mission, vision, and implementation of projects for the Department of Defense to align current and future strategic plans and objectives. She possesses proven expertise in both program management and clinical expertise in research, business development, and wounded care. Proud spouse and teammate to Lieutenant Colonel Jeff Scott, and mom to Evan and Quinn.

Don Philpott is editor of *International Homeland Security Journal* and has been writing, reporting, and broadcasting on international events, trouble spots, and major news stories for almost forty years. For twenty years he was a senior correspondent with Press Association—Reuters and the wire service, and traveled the world on assignments including Northern Ireland, Lebanon, Israel, South Africa, and Asia.

He writes for magazines and newspapers in the United States and Europe and is a regular contributor to radio and television programs on security and other issues. He is the author of more than 140 books on a wide range of subjects and has had more than 5,000 articles printed in publications around the world. His recent books include the Military Life series, *Terror—Is America Safe?*, *Workplace Violence Prevention*, and the *Education Facility Security Handbook*.